# Food, Globalization and Sustainability

# Food, Globalization and Sustainability

*Peter Oosterveer and David A. Sonnenfeld*

London • New York

First published 2012
by Earthscan
2 Park Square, Milton Park, Abingdon, Oxon OX14 4RN

Simultaneously published in the USA and Canada
by Earthscan
711 Third Avenue, New York, NY 10017

*Earthscan is an imprint of the Taylor & Francis Group, an informa business*



*British Library Cataloguing in Publication Data*
A catalogue record for this book is available from the British Library

*Library of Congress Cataloging in Publication Data*
A catalog record for this book has been applied for

ISBN: 978-1-84971-260-6 (hbk)
ISBN: 978-1-84971-261-3 (pbk)

Typeset in Sabon and Frutiger
by MapSet Ltd, Gateshead, UK

Printed and bound in Great Britain by the MPG Books Group

# Contents

# List of Figures, Tables and Boxes

## Figures



## Tables

## Boxes

# List of Acronyms and Abbreviations

| | |
|---|---|
| AIDCP | Agreement on the International Dolphin Conservation Program |
| ARC | Agricultural and Rural Convention |
| ASC | Aquacultural Stewardship Council |
| BSE | bovine spongiform encephalopathy |
| BSI | British Standards Institution |
| CAFO | confined animal feeding operation |
| CAP | Common Agricultural Policy |
| CBD | Convention on Biodiversity |
| $CO_2e$ | $CO_2$-equivalent |
| COP | Conference of the Parties |
| CSA | community-supported agriculture |
| CSR | corporate social responsibility |
| DDT | dichlorodiphenyltrichloroethane |
| DEFRA | Department for Environment, Food and Rural Affairs (UK) |
| EEZ | exclusive economic zone |
| EFSA | European Food Safety Authority |
| EII | Earth Island Institute |
| EU | European Union |
| EUREP | Euro-Retailer Produce Working Group |
| FAO | Food and Agriculture Organization of the United Nations |
| FLO | Fairtrade Labelling Organizations International |
| FoEI | Friends of the Earth International |
| FOS | Friends of the Sea |
| GATT | General Agreement on Tariffs and Trade |
| GCC | global commodity chain |
| GDP | gross domestic product |
| gha | global hectare |
| gha/cap | global hectares per capita |
| GHG | greenhouse gas |
| GlobalGAP | Global Good Agricultural Practices |
| GM | genetically modified |
| GMO | genetically modified organism |

| | |
|---|---|
| GPN | global production network |
| GRI | Global Reporting Initiative |
| HACCP | Hazard Analysis and Critical Control Point |
| ICFFA | International Commission on the Future of Food and Agriculture |
| IDRC | International Development Resource Centre |
| IFOAM | International Federation of Organic Agriculture Movements |
| IFPRI | International Food Policy Research Institute |
| ILO | International Labour Organization |
| IMF | International Monetary Fund |
| IPCC | Intergovernmental Panel on Climate Change |
| IPPC | International Plant Protection Commission |
| ISO | International Organization for Standardization |
| ISRTA | International Standards for Responsible Tilapia Aquaculture |
| LCA | life-cycle analysis |
| MELJ | Marine Eco-Label Japan |
| MJ/kg | megajoule per kilogram |
| MSC | Marine Stewardship Council |
| NAFTA | North American Free Trade Association |
| NGO | non-governmental organization |
| NOGAMU | National Organic Agricultural Movement of Uganda |
| OECD | Organisation for Economic Co-operation and Development |
| OIE | International Office of Epizootics |
| PAS | Publicly Available Specification |
| PDO | protected designations of origin |
| PGI | protected geographical indications |
| ppm | process and production method |
| RFMO | regional fisheries management organization |
| SAN | Sustainable Agricultural Network |
| SCM | Subsidies and Countervailing Measures |
| SPS | Sanitary and Phytosanitary |
| TBT | Technical Barriers to Trade |
| TRIPS | Trade Related Aspects of Intellectual Property Rights |
| UK | United Kingdom |
| UN | United Nations |
| UNCED | United Nations Conference on Environment and Development |
| UNCLOS | United Nations Convention on the Law of the Sea |
| UNDP | United Nations Development Programme |
| UNFCCC | United Nations Framework Convention on Climate Change |
| US | United States |
| USDA | United States Department of Agriculture |
| WHO | World Health Organization |
| WSSD | World Summit on Sustainable Development (United Nations) |
| WTO | World Trade Organization |
| WWF | World Wide Fund for Nature |

# Preface

When Tim Hardwick, editor at Earthscan, suggested that I write a textbook on food, globalization and sustainability, I immediately thought of inviting David A. Sonnenfeld to this project. Producing a textbook on such a vast subject requires the writers to make choices and I knew David would make sure that these would be balanced and well thought through. We are very glad that in this way writing the book became a shared project. Despite the inevitable choices in content that had to be made, we believe that readers will find this volume an up-to-date introduction to the global world of food. Our aim has been to show how the social sciences can clarify some of the key global sustainability challenges that contemporary food provision is faced with. Ultimately, we hope that this book contributes to greater sustainability and justice in food production and consumption around the world.

Writing this book would not have been possible without contributions from many people. First, we would like to acknowledge the many students who have participated annually since 2002 in the course on globalization and sustainability of food production and consumption at Wageningen University, the Netherlands. Other lecturers in this course were Alberto Arce, Henk Renting and Han Wiskerke; we want to thank them for their collaboration. Most of the subjects presented in this book have been discussed at numerous conferences, seminars and workshops organized on nearly all continents. We have profited much from the debates during these meetings.

Next we want to thank Donna Maurer for her conscientious language editing and her flexibility in adapting to our tempo; and the staff at Earthscan: Tim Hardwick (editor) for his support and patience, Claire Lamont (assistant editor) and Laura Briggs (publications coordinator) for their help. We also want to thank all colleagues at the Environmental Policy Group at Wageningen University who contributed directly and indirectly to this book project. Thanks to Suzanne van der Schenk for her contribution in designing the cover and for all moral support. We also want to thank our families for the patience they showed during the writing of this book.

We hope readers will enjoy this book as a contribution to understanding how what we eat can be more sustainable and globally just and how future generations can have access to enough and healthy food.

*Peter and David*
*Wageningen, the Netherlands, and Syracuse, New York*
*March 2011*

# 1

# Introduction

Food is vital for human life. Since we all must eat, food should be foremost on everyone's agenda. This is not the case for many, however, who take the availability of sufficient quantities of healthy food for granted. Yet one in seven individuals worldwide – nearly 1 billion people – go hungry, and concerns are growing about whether the world will produce enough food for the future.

Even at the time of this writing, the Food and Agriculture Organization of the United Nations (FAO) is warning about rapid rises in global food prices. The trends of steadily increasing food production and long-term declining prices seem to be reversing. After the price hike in 2008, figures point to new record prices in 2011, which may exacerbate food security problems for hundreds of millions of poor people. Future food supply cannot be taken for granted; many governments have ignored food policy for too long. They have shown much complacency about food provision and are increasingly reluctant to engage the problem seriously. In February 2011, *The Economist* included a special report on the future of food, arguing that the world is facing a food system in crisis. In addition to temporary problems, the report's authors also observe structural problems (*Economist*, 2011). Food production will have to rise by 70 per cent by 2050 to keep up with a global population growing to 9 billion, with the explosion of megacities in developing countries and with changing diets in such countries as China and India. The necessary increases in food production will have to be realized in more challenging conditions than in the past, as there is little unfarmed land remaining, and less water available, while proven recipes of the past (using high-yielding crop varieties in combination with more intensive use of fertilizer and pesticides) are no longer applicable in many parts of the world. These problems are exacerbated by climate change and the global reduction of biodiversity. These observations clearly support putting food at the centre of public interest. We need to better understand global food provision, its sustainability and its future. This book is intended to contribute to achieving this aim. In the next 11 chapters, we explain how contemporary food provision is changing in the context of globalization, illustrate especially serious sustainability problems and possible responses and discuss future perspectives towards greater sustainability.

The book examines these fundamental changes in contemporary food provision and the new challenges resulting from them, taking an environmental social science perspective. In this chapter, we give a brief overview of the book, explain the perspective from which we approach the world of food and introduce the different chapters.

## Background

The world of food is changing radically. Very visible is the enormous technological capacity for producing and processing food that is available today as compared with the past. But we also know much more about the possible adverse consequences of using these modern technological options. Next, although national food policies remain to be developed and implemented, they are increasingly coordinated with other national governments. We can see the growing influence of multilateral institutions, such as the World Trade Organization (WTO), on national policies, and sometimes even private industries have a large influence on food policies. Globalization has had a big impact on these fundamental changes taking place in contemporary food provision. In our globalizing world, people's lives are determined not only by their particular local conditions but also increasingly by distant developments. Someone's actions can have consequences at large distances in time and space. Through globalization, food today is increasingly traded internationally, transforming its production and consumption patterns worldwide and influencing many food-related practices. The world of food that used to be dominated by farming and rural dynamics has become more influenced by consumption and dynamics in retail. These changes generate new challenges, such as how to increase sustainability in food provision, reduce the negative social impacts of international trade and govern food from a global perspective. Or, as the United Nations' (UN) special rapporteur Olivier De Schutter mentions in his report to the UN Human Rights Council (De Schutter, 2010, p1), reinvesting in agriculture is essential to the concrete realization of the right to food, but in the context of ecological, food and energy crises, the most pressing issue 'is not how much, but how'.

These fundamental transformations and emerging challenges call for more attention to food in general and more research in particular, but they also require reconsidering the concepts that we conventionally use to analyse them. Many concepts that were probably adequate in the past are no longer appropriate to analyse current dynamics. To understand contemporary food provision, we therefore need different conceptual frameworks because otherwise we may not be able to respond to the emerging problems. This book therefore builds on a tradition of analysing food provision from a social-science perspective but tries also to critically assess the applicability of conventional conceptual frameworks. In the (food) social-science tradition, farming has remained the focus of attention for a long time, but in recent decades this focus has been complemented with studies on processing, retailing and consumption. We intend to contribute to this movement towards a more integrated approach to analysing food provision in its entirety and therefore focus on analysing sustainable food provision under the conditions of global modernity. We pay particular attention to the changing relationships between food producers and consumers and the essential roles played by retailers and non-governmental organizations (NGOs). In doing so, we apply a broad social-science perspective on food provision and, where appropriate, use findings from the natural sciences to better understand relevant human behaviours.

In this book, we aim to contribute to essential debates about the future of food from a global perspective. It is not easy to do justice to all the different views and experiences around the world, biased as we are by our own backgrounds and experiences. Nevertheless, we bring in many examples from Africa and Asia, as well as others from Western Europe and the United States (US). Through these contributions, we make an effort to present a balanced overview of the developments and related debates currently taking place on globalizing food provision.

## Overview

The book is organized in three main sections, supplemented by this introduction and a concluding chapter. The first section presents several key conceptual tools for analysing globalization and its relationship to the sustainability of contemporary food provision. The second section offers several empirical cases illustrating efforts to address sustainability in the global food supply. The third main section discusses the roles that growers, consumers, food retailers and others can play in contributing to greater sustainability in global food provisioning.

### Section I: Conceptual background

Globalization, sustainability and changing notions of the need to regulate food are the focuses, respectively, of the three chapters making up Section I of this volume.

Globalization of food production and consumption is the subject of Chapter 2. This chapter starts with several empirical observations, such as the substantial increase of international food trade and the high impacts that the process of globalization has on the organization of food production and consumption, including on the roles of national governments in its regulation. We subsequently review several conceptual frameworks that have been developed to better understand these changes. Theories on global commodity chains, global production networks, convention theory and global networks and flows are presented and discussed. The chapter concludes by underlining the importance of bringing together the global and the local dynamics within the framework of networks and flows. This conceptual framework highlights the growing tensions between global and local dynamics of food and the reasons why innovative governance arrangements are needed for more sustainability in globalizing food provision.

The concept of sustainability is the subject of Chapter 3, in which we show the ways in which many view modern agriculture as an important cause of environmental problems. Modern, intensive farming methods have direct impacts on natural ecosystems and human health, and this had already been discovered by the 1960s. Since then, various other food-related environmental concerns, such as worries about animal welfare, food safety, energy use, landscape, climate change and biodiversity, have surfaced in public debates. Nowadays, sustainability concerns related to food production and consumption entail, among others, the use of non-renewable (fossil fuels, phosphate) and renewable (solar and wind energy, water) natural resources, the impacts on atmosphere and climate (greenhouse gas (GHG) emissions), soil fertility and land and water management, (agro)biodiversity, pesticides use, animal welfare, waste disposal, economic practices and environmental policies. These concerns are framed within a wider understanding of sustainability, where the existence of multiple definitions leads to much confusion on possible strategies. The context of globalization again changes our understanding of sustainability and opens up new debates on potential tools that could be applied to increase the sustainability of contemporary food provisioning.

This conclusion provides the starting point for Chapter 4, which is devoted to the regulation of food in the global network society. Conventionally, regulating food was based on decisions by autonomous sovereign national governments to develop their economies and to feed their populations by securing access to sufficient and safe food. Nowadays, these authorities are still sovereign, but their autonomy in making policy decisions has drastically declined. Their capacities to effectively control food have diminished because international trade is rapidly expanding. Governments are faced with increasing volumes of imported food, whereby private actors are occupying key controlling positions. At the same time, the demand for regulating food provision is growing, as the sustainability of food is raising public concern, and such concern is particularly invigorated by the introduction of innovative technologies with unknown consequences for human and environmental health.[1] Today consumers are concerned not only about the quality, safety and price of their food but also about the health, social, ecological and animal welfare impacts occurring at all different stages of the supply chain. Therefore, food regulation must be organized differently in the context of global modernity. In reaction, innovative arrangements have been developed, including multilateral agreements and institutions, private standards and private labelling and certification schemes. This chapter presents key examples of these different arrangements, including their background and impacts.

## Section II: Case studies

Building on the conceptual perspectives introduced in the first section, Section II of this volume presents four case studies on important dimensions and dynamics of globalization, food and sustainability.

Chapter 5 presents the case of climate change, as this is one of the most urgent global environmental problems the world faces today, with high relevance for global food provision. Rising temperatures and sea levels and increasingly volatile weather conditions may harm the livelihoods of many people and threaten future food provision. At the same time, food production, processing, transport and consumption are also important contributors to the problem of global warming. This chapter presents the background of the relationships between global warming and food provision and presents some key indicators and measurement tools. A global consensus on the causes of climate change seems to be growing, but much controversy remains on what actions need to be taken to prevent further degradation. In this chapter, we discuss several competing views on the relation between global warming and the globalization of food provision and the different strategies that emerge from them. This results in an overview of the different strategies, management tools and governance arrangements that are available to mitigate the impact of food provisioning on climate change.

Chapter 6 illustrates how an alternative movement is promoting local agri-food networks as a response to contemporary industrialized food provision. A variety of local food initiatives is rapidly acquiring shape where different social actors concerned about contemporary food find each other. Small farmers, (urban) consumers, local (food) retailers, local governments and business associations, and social activists seeking to boost local economies and to improve the quality and sustainability of their food find common interests in these initiatives. Different social actors work together to support local agricultural production and consumption and create alternative, short food-supply chains. Through these efforts, they try to provide a counterpoint to the overall trend towards increasingly globalized food provision. They argue that the awareness and social bonds necessary to strengthen sustainability in its social, economic and environmental aspects can be recreated through directly connecting food producers, consumers, retailers and relevant institutions. This chapter presents the background of these initiatives and illustrates their diversity in aims, organizational set-ups and impacts. Some alternative, local food networks try to optimize the environmental impacts of food by demanding minimal energy for transport, processing and packaging, and maximizing freshness and quality. Others aim primarily to support local communities by forming knowledgeable relationships between food producers and consumers, supporting local farmers and supplying more unprocessed and seasonal foods. Improving school meals is an example of this strategy. Yet other local networks try to improve local food security for the urban poor. This chapter also examines critiques of such alternative food-supply strategies, including whether the latter effectively address problems of sustainability and if they can provide a viable alternative to steadily globalizing agri-food production and consumption.

Next, Chapter 7 uses fair trade to illustrate the innovative networks that emerge between food producers and consumers in global modernity. Fair trade already has a history of some 70 years, but in the last 25 years, it has grown into an exemplary alternative standard for trade arrangements in international markets involving developing countries. Fair trade intends to support smallholder producers in developing countries by increasing their share of the commodity price and to secure their position on the global market. In this chapter, we introduce fair trade and its formal aspects, such as the labelling arrangements and the relevant institutions involved. International coffee trade was the first case where fair trade's aims were realized, so this example is used to show how fair trade operates in practice and its impacts on both producers and consumers. In this chapter, we also discuss the main criticisms of the principles and the impact of fair trade to assess its future perspectives as an innovative relationship between producers and consumers and whether fair trade can contribute to more sustainable global food provision.

Chapter 8 uses fish provision as a case to illustrate the global character of present problems associated with sustainability. Contemporary fisheries are faced with depleting (or even collapsing) fish stocks, threatening their future as a source of food. The case of global fish clearly illustrates the dynamics involved in contemporary food provision, whereby sustainability aspects cannot be ignored. Fish provision has become embedded in global networks and flows involving multiple actors, including private corporations, NGOs and retailers, often operating at large distances from each other. At the same time, the capture and farming of fish remain largely dependent on specific local ecological and climatic conditions. Thus global fish trade and local fisheries are inextricably and irreversibly bounded through dynamic relationships. In this chapter, we summarize the present state of global fish provision, paying attention to both capture fisheries and aquaculture. This description provides ample background for an analysis of the key sustainability challenges in fisheries management and fish farming (aquaculture). Under these conditions, imposing internationally coordinated mandatory regulations proves complicated, because it requires unanimous support from all governments involved, which is hard to achieve. At this moment, different private governance arrangements are being introduced, and we present the Marine Stewardship Council (MSC) as an innovative certification and labelling scheme, and consumer guides as a consumer-oriented environmental policy instrument. We then focus on how these initiatives understand sustainability, as well as the roles of different stakeholders in these arrangements.

## Section III: Future perspectives

Section III examines the future of food, which we consider through chapters on the roles of producers, retailers and consumers in promoting the sustainability of food in the context of globalizing modernity.

Chapter 9 deals with the changing roles of the primary food producers: farmers. Nowadays, farmers find themselves operating in a rapidly changing environment wherein they are faced with multiple challenges, including how to secure the future of their farm, their livelihood and their local community, but also how to contribute to more sustainable food provision. This encompassing challenge cannot be treated in its entirety in one chapter, so we focus on the different strategies that farmers develop in response to the demands they are facing. Some farmers focus on high-tech methods as a means to reduce environmental impacts, while others try to embed their production practices in local communities and local ecosystems with the same intention. Each of these strategies incites extensive scientific and public debates on its consequences for globalization and sustainability in farming and food for the future. After presenting the various farmer strategies, we review the debates on genetic modification, regional food labels and multi-functionality, and finally, organic agriculture. The perspectives of small farmers receive special attention, as they may face certain problems when determining their preferred strategy.

In the next chapter, Chapter 10, we consider the (future) role of retailers in global food provision. Supermarkets have recently become central locations for selling and buying food, not only in richer countries, but increasingly in many other parts of the world as well. As the obligatory passage points for most food sales, supermarkets are essential to the understanding of contemporary global food provision. Supermarkets have a profound impact on all stages of the food supply chain, including farming, processing, transport, trade and consumption, and thus they have acquired a central coordinating position in food supply chains. This chapter briefly reviews the main societal trends that promote the spread of supermarkets, including rising incomes, urbanization, increasing female participation in the labour force and a desire to emulate Western culture. Then we analyse the different roles supermarkets can occupy in contemporary food provision. This analysis provides the starting point for a review of supermarket strategies in different countries and will enable us to identify their prime contributions to promoting sustainability in food provision, in comparison with other social actors, including governments.

Chapter 11 is the last chapter in this section and considers the involvement of consumers in sustainable food provision. For a long time, consumers were ignored as relevant stakeholders in promoting sustainable food provision, because they were mostly considered unreliable and unmanageable. This impression has changed in recent years, because the role of consumers has become quite influential in debates on the future of food. To assess consumers' changing position, however, it is important to have a better understanding of their complicated behaviours. This chapter therefore reviews the relevant conceptual debates within the social sciences on how to analyse consumers and their behaviours. From these debates, we conclude that a social-practices perspective on consumer behaviour is most promising, because it allows for combining more structuralist perspectives with an understanding of people as being actively involved in shaping their behaviours as consumers. When applying this social-practices perspective on consuming food, its routinized character clearly stands out. Food consumption is embedded in everyday life and represents a collection of daily routines, culturally embedded practices and economic calculations. It entails a broad array of activities, such as buying products and services, transporting, preparing and eating them, and finally disposing of or recycling the remaining waste. These food consumption practices are not very amenable to changes, such as buying more sustainable food, despite the positive attitudes that many consumers express towards this goal. With the help of the social-practices perspective, it becomes possible to reflect more meaningfully on the possible roles consumers can play in promoting sustainable food provision in the future. Changing consumer behaviours can be linked to cultural change, trust, provider strategies and changing biographies. The cases of organic food consumption and dietary change are used to illustrate these dynamics more concretely.

### Concluding chapter

The final chapter of this book, Chapter 12, summarizes important changes taking place in global food production and consumption; considers some of the different public and scientific debates and agendas that are being advanced related to the future of food provision; and discusses how globalization and sustainability figure into these debates and agendas. Specific attention is given to how food policy and technologies are embedded in social and environmental dynamics. As with the entire volume, we try to explain important contributions that a broad (environmental) social-science perspective can offer to identifying pathways through which greater sustainability can be achieved in producing, trading, processing and consuming food within global modernity, with various social actors enacting distinct roles.

## Conclusion

There are many reasons to closely consider the challenges of contemporary food provision, its relation to accelerated globalization and requirements for greater sustainability. Paramount among these is the serious possibility that food availability may worsen for many. Volatile prices, lagging production, population growth and strains from climate change all contribute to the immensity of these challenges. With the ultimate aim of helping find ways to secure sufficient quantities of nutritious food for all in coming decades, this book offers students and scholars a broad understanding of the main issues at stake and suggests ways to move forward towards a more sustainable and fair provision of food globally.

## Note

1 For instance, in December 2010, German authorities discovered that a feed fat-producing company had mixed fat intended for animal feed with a batch of fatty acids for technical purposes, contaminated with dioxins. In response, 1100 farms were shut down because eggs and meat containing dioxins had been discovered. For many weeks German consumers refused to buy eggs (EC, 2010, 2011).

## References

De Schutter, O. (2010) *Report Submitted by the Special Rapporteur on the Right to Food*, United Nations General Assembly, Geneva

EC (European Commission) (2010) *RASFF Alert*, 2010.1771, European Commission, Brussels

EC (2011) *Food Navigator*, 6 January, European Commission, Brussels

*Economist* (2011) 'The 9 billion-people question: A special report on feeding the world', *The Economist*, February, vol 398, no 8722

Section I

# Conceptual Background

# 2

# Globalization and Food Production and Consumption

This chapter aims to:

- introduce the concept of globalization;
- present the main trends in the globalization of food provision;
- discuss the most relevant social-science perspectives on globalization and food.

## Introduction

International trade in food products has been increasing for many years. And, although most food produced around the world is still consumed domestically, the process of globalization deeply influences the organization and sustainability of food production and consumption worldwide. Globalization has substantial impacts on the changing roles that national governments, private companies (producers, traders, processors and retailers) and consumers play in contemporary food provision.

This chapter first provides some factual background to this globalization process and then proceeds by reviewing several conceptual frameworks that have been developed to better understand these changes as they relate to agriculture and food provision. We discuss the ideas of global commodity chains, global production networks, convention theory and global food flows. We then summarize important debates in global food provision, in particular, food security, sustainability and (global) food governance.

The chapter concludes by emphasizing the importance of addressing the tension between the emerging dynamics of the global 'space of flows' and the local 'space of places', because here we find reasons for environmental problems related to contemporary food provision. With a better understanding of these dynamics, it becomes possible to analyse the background of such problems and to identify possible ways of moving forward, including innovative governance arrangements that may offer better solutions than those currently in place. These issues are discussed further in subsequent chapters of this book.

# Globalization

## What is globalization?

The modernization process has entered a new phase since the 1970s, when the states that were essentially oriented towards organizing their respective national societies internally, for instance by managing them as welfare states, were forced to consider international developments more explicitly. This new phase is commonly designated as the process of accelerated globalization, whereby internationally operating companies become dominant and the exchange of products, finances and information increasingly become organized on a worldwide scale. Globalization is the ongoing process by which regional economies, societies and cultures become integrated through a worldwide network of exchanges of material goods, people, ideas and information (Robbins et al, 2010). Some consider this to be an ongoing process towards further homogenization, as exemplified in the area of food by the global spread of fast-food restaurants (Ritzer, 1996), where people around the world are becoming part of one uniform eating culture. Others contest this observation, taking the view that variations between countries and people persist and that a growing diversity results from translating global phenomena into local contexts (Counihan and Esterik, 1997; Watson and Caldwell, 2005). They point to the importance of not only looking at globalization from an economic perspective, but also to pay attention to social and cultural dynamics. For a more elaborate discussion, see Held et al (1999) and Mol (2001).

The global is not to be situated as simply opposing the local because, as Massey (2004) and Born and Purcell (2006) have stressed, the 'local' and the 'global' are not two separate worlds. Space must be thought of relationally, such that 'global space' is the sum of relations, connections, embodiments and practices, constituted and negotiated through '(local) places' rather than a separate spatial sphere. Scale has a socially and politically constructed character, and thus globalization implies transformations of a range of mutually related social practices and institutions at different levels of scale and not simply the scaling up of social and economic life or the rapid increase of worldwide trade. Global and local are inextricably and irreversibly bound together through dynamic relationships, whereby connections between the two may be more or less mobile, more or less intense, more or less social and more or less 'at a distance' (Beck, 1997).

The definition of globalization nevertheless remains contested, as few definitions seem to comprise all relevant economic, political, social and cultural dimensions. Among all dimensions of globalization, we consider it important to highlight its flexibility and the dynamics that prevent it from becoming a one-sided process of increasing dominance from the centre to the rest of the world. Therefore, we consider Giddens' (1990, p64) definition to be the most appropriate: 'the intensification of worldwide social relations which link distant localities in such a way that local happenings are shaped by events occurring many miles away and vice versa'.

## How is food related to globalization?

Agriculture and food constitute one domain in society where globalization is clearly taking place, as food is increasingly produced, traded and consumed in ways that display worldwide dynamics. Large quantities of food are traded internationally and in 2009 agricultural exports represented a total value of US$1169 billion, or 9.6 per cent of total global trade (WTO, 2010). Agricultural trade has grown substantially over the years, despite regular decreases because of economic crises and natural disasters (see Table 2.1).

Despite substantial growth, global agricultural trade has increased at a much slower pace than trade in other products. Prior to the 1960s, agricultural products accounted for more than 30 per cent of all goods traded globally, while in the early 2000s this declined to less than 9 per cent (Anderson, 2010). Rich countries in the West and emerging economies such as China and Brazil dominate agricultural trade, while African food exports have substantially declined in recent decades (see Table 2.2).

Despite this growing trade in agricultural products, most food is still consumed domestically and global food trade continues to represent only a small percentage of total production. Table 2.3 shows that, for most products, less than 12 per cent of total production is exported. These figures are only substantially higher for tropical products such as tea and coffee, for which more than 80 per cent of the total production is exported.

Another example of the continued importance of domestic production and consumption is that the share (in volume) of imported food in the US was 15 per cent in 2005, up from 12 per cent in 1990; and thus 85 per cent was still domestically produced.

Even though food exports represent only a minor share of global agricultural production, the organization of food production and consumption is transforming globally, and most food-related practices are influenced by it, as exemplified by the growing trade in fresh fruits and vegetables (see Box 2.1).

**Table 2.1** *Annual changes in world trade in agricultural products (1980–2009)*

| *Years* | *Annual percentage change* |
|---|---|
| 1980–1985 | −2 |
| 1985–1990 | 9 |
| 1990–1995 | 7 |
| 1995–2000 | −1 |
| 2000–2009 | 9 |

*Source:* WTO (2010)

**Table 2.2** *Leading agricultural exporters and importers (relative shares in 2008)*

| *Country* | *Export (%)* | *Country* | *Import (%)* |
|---|---|---|---|
| US | 16 | European Union (27 member states) | 18 |
| European Union (27 member states) | 14 | US | 12 |
| Brazil | 7 | China | 9 |
| Canada | 6 | Japan | 8 |
| China | 5 | Russian Federation | 4 |
| Argentina | 4 | Canada | 3 |
| Indonesia | 4 | Republic of Korea | 3 |
| Thailand | 3 | Mexico | 3 |
| Malaysia | 3 | Others | 42 |
| Australia | 3 | | |
| Russian Federation | 3 | | |
| Others | 33 | | |

*Source:* RaboBank Nederland (2010)

Food provision is becoming increasingly interdependent worldwide. Food consumption, production and marketing practices are influenced more by global forces, such as demographic, economic, political and environmental developments, than by local market conditions. For instance, hunger still has not vanished from our planet. Local weather and climate crises and political instability offer a partial explanation for the lingering hunger problem, but the 2008 food crisis, in particular, definitely had global roots. In addition to failed harvests in Australia and declining stocks, speculation on the futures markets and the search for stocks to produce biofuels contributed as well.

Another example is the pressure to create liberal trade regimes that open up opportunities for large corporations to profit from the shifting distribution of supply and demand. Today, several large companies dominate food trade and processing worldwide (Bonanno et al, 1994). For example, only ten companies control 50 per cent of the global seed market, while in the global pesticides market, the ten largest companies control 82 per cent. In the global food market, the top ten companies have a 28 per cent share (Dalle Mulle and Ruppanner, 2010).

**Table 2.3** *Global production and trade (million tonnes for 2009/2010)*

| *Product* | *Production* | *Trade* | *% Exported* |
|---|---|---|---|
| Wheat | 682.6 | 128.1 | 18.8 |
| Coarse grains | 1125.2 | 114.7 | 10.2 |
| Rice | 455.6 | 30.8 | 6.8 |
| Cassava | 251.0 | 28.2 | 11.2 |
| Sugar | 156.7 | 53.3 | 34.0 |
| Meat | 283.9 | 25.4 | 8.9 |
| Dairy | 698.8 | 43.5 | 6.2 |
| Fish | 145.1 | 54.9 | 37.8 |

*Source:* FAO (2010a)

### Box 2.1 Global trade in fresh fruits and vegetables

Fresh fruits and vegetables constitute a category of agricultural products for which trade has grown substantially in just a few years. The value of international exports in these products increased from $71 billion in 2001 to $152 billion in 2009.[1] This highly dynamic sector is profiting from rising incomes and health concerns among consumers in the West, in combination with technological improvements that allow for better transport over longer distances. Developing countries are profiting from these increased opportunities and increasing their shares in the global trade of fresh fruits and vegetables substantially.

*Source:* Diop and Jaffee (2005)

Also, food regulation today is taking place less at the national level and increasingly at the global, local and regional levels. Food safety requirements in the European Union (EU) have worldwide impacts, and agreements on agriculture concluded within the WTO are binding for all of its 153 member states. The conventional role of national governments in agriculture and food politics is changing as well, and much more is determined by producers, traders and processors, as well as the consumers themselves.

These developments require more in-depth analysis for a better understanding of their differential impacts on developing countries, on smallholder farmers and on the environment. Towards this end, several conceptual frameworks on globalization and food have been introduced. Next, we present and discuss some of the most relevant ones.

## Key Conceptual Frameworks on Globalization and Food

### The political economy of food production and consumption

Scholars operating within the framework of political economy base their analyses on the relationship between political and economic dynamics, and aim to demonstrate how the globalization of agriculture and the food industry is proceeding through examining shifting balances of power (Busch and Juska, 1997; Bair, 2009). They depict contemporary food provision as an economic system where agricultural commodities are produced in 'peripheral' regions of the global economy for retail and consumption in the 'core' regions of the US, Europe and Japan. At the same time, processed food products are exported from core countries and compete under unfair conditions in non-core countries. Using ideas developed by Marx and applying adaptations proposed by Wallerstein (1974) when analysing the modern global capitalist economy, these political economists account for the transformation and industrialization of capitalist agriculture, the rising power of multinational food and agribusiness corporations and the global integration of the agri-food system. They consider this to be a long-term historical trend that has existed since the colonial era of the 16th and 17th centuries. As Friedmann (1995, p18) argues: 'The world has been disrupted and integrated on a global scale for centuries, and people, plants, and practices have been relocated and reshaped many times.' Friedmann points to the search for spices from Asia, whereby Europeans conquered much of the globe and forced or induced labourers on plantations and farms to produce sugar, coffee, cocoa, tea and opium to stimulate, soothe and compensate workers concentrated in factories and cities by the industrial transformation of life in Europe.

Friedmann and other scholars argue that analysing contemporary agriculture and food production from the perspective of a single country is futile and that a global perspective is necessary. From the 19th century onwards, former settler colonies, including the US, Australia and New Zealand, provided food for the workers in the industrializing European countries, while developing an industrial sector themselves. These scholars designate this particular agri-food complex as a 'food regime', which 'links international relations of food production and consumption to accumulation broadly distinguishing periods of capitalist transformation since 1870' (Friedmann and McMichael, 1989, p95). They identify two main food regimes: the first regime unfolding between 1870 and 1914, and a second following World War II, since 1945. In the first food regime, 'settler agriculture provided exports of dietary staples (wheat, meat) as essential wage foods, underwrote industrial profits by lowering food production costs, and absorbed surplus labour from the European countryside through international migration' (Friedmann and McMichael, 1989, p111). In the second food regime, agriculture increasingly became tied to industrial capital and the modern state system. In the latter period, state regulation underpinned a new and inherently national form of accumulation based on high wages and mass consumption of cheap (standardized) food products.

During the 1970s and 1980s, however, the growth of transnational corporations meant a shift from state to private (international) capital as the dominant structuring force. Some researchers therefore suggest that there has been a transition to a third food regime since the 1980s. The second food regime ran into a crisis because of overproduction, leading to price instability and increased competition on export markets. The surpluses produced from agriculture in the EU and US promoted a politics of neoliberalization in the food sector. The era of uniform, cheap mass products seemed to have passed and the balance of power shifted away from the producing side of the food supply chain to the retailing and consumption side. This emerging third food regime, sometimes called 'post-productivism', is characterized by increased flexibility and diversity in the food products available on the market and integration along global food supply chains unconstrained by national boundaries, while the nation state became fragmented and no longer the undisputed coordinator of national economies. The logic of the free market (the 'Washington consensus' and the neoliberal project) guided globalization as a worldwide development project. Technological progress and globalization have turned simple national or regional food systems into large and complex international agri-food chains. Friedland (2005) describes globalization in food as also facilitated by the concentration of food retailing in metropolitan markets, and driven by the expectations of consumers in developed countries for the year-round availability of fresh produce, including a variety of tropical products. In reaction to this global corporate food regime, counter-movements emerged, expressing a discourse of diversity, local ownership, food safety and security (Ilbery, 2001). Examples of this are farmers' and citizens' organizations promoting organic farming, modes of traditional farming, food sovereignty, community-supported agriculture (CSA) and fair trade.

Agriculture, no longer the embedded localized means of survival it was in the past, today has become integrated in globalizing supply chains. McMichael (2000, p23)[2] concludes: 'under these conditions, which affect world regions differently, agriculture becomes less an anchor of societies, states, and cultures, and more and more a tenuous component of corporate global sourcing strategies'. According to the Marxist terminology, agriculture and food have become increasingly 'commoditized', which refers to the growing emphasis on the price and market value of agricultural products, and to the diminished attention given to those products' use value (such as nutritional quality). Commoditization has promoted the role of markets in selling and buying agricultural products, and a growing volume of agri-food products is circulated globally, in ways comparable to those of other commodities, such as cars and computers. Food, however, still has a distinctive character (i.e. its organic character) compared with most other commodities. Agriculture is the social organization of a biological system, and as a consequence, not all agricultural products and not all components of agri-food systems can be globalized in the same manner. Every agri-food product has a distinctive history of uneven and combined development that is 'twinned with diversity in commodity regulation and organization' (Friedland, 2005, p28). The particular characteristics of each food product are captured in the notion of agri-food 'commodity chains', which link the different phases a food product goes through while becoming a commodity and how these phases are managed. See Box 2.2 for an example of a commodity chain analysis.

With the help of the global commodity chain (GCC) concept, scholars attempt to examine the different ways in which production and distribution activities are integrated. A GCC is a particular type of value chain, which is the 'process by which technology is combined with material and labour inputs, and then processed inputs are assembled, marketed, and distributed' (Gereffi et al, 2005, p79). Scholars applying a GCC perspective focus on such questions as: who controls global trade and industry? How do they exercise this control? And what are the consequences for farmers, particularly those in developing countries? They intend to explain the growth in global food chains by analysing the link between the rise of a specific group of economic agents and the expansion of globally dispersed trade-based production networks. In GCCs, brand owners and large supermarkets exercise power through their central role in system coordination (Daviron and Gibbon, 2002). Global agri-food commodity chains are considered as being driven by buyers, in contrast with, for example,

### Box 2.2 A commodity chain analysis of export horticulture in Brazil

In the São Francisco Valley of northeast Brazil, the irrigated area expanded rapidly after 1960 because of large-scale state investments. Later, state-financed development agencies settled farmers to produce basic food crops for the domestic market, but by the 1980s these agencies shifted to promoting export crops to earn the foreign exchange Brazil urgently needed. By 2001, the valley produced relatively high-value fruits, such as grapes, with 1400 farms and around 20,000 workers. Medium-sized producers were supported in these efforts by collective action and public assistance. The developing international commodity chain was assisted by a large São Paulo-based co-operative with extensive experience in exporting to Europe. By the early 2000s, buyer and importer power was enhanced, and European retailers imposed strict requirements on safety and quality. To secure food safety, famers had to be certified according to Global-GAP (Global Good Agricultural Practices) standards, while packing houses had to have HACCP (Hazard Analysis and Critical Control Point) certification.[3]

Producers were supported by different local institutions and several actors from within the grape commodity chain, in particular, importers. Importers sought to strengthen their position by organizing the producer end of the chain to provide retailers with high-quality fruits. Analysing the interactions between farmers' institutions and private companies within the commodity chain illuminates how such chains can maintain their competitiveness.

*Source:* Selwyn (2008)

the global automobile industry, which is much more driven by producers (Gereffi et al, 2005). The concept of GCCs can be used to analyse the:

> *full range of activities* including coordination *that are required to bring a specific product from its conception to its end use and beyond. This includes activities such as design, production, marketing, distribution, support to the final consumer, and governance of this entire process.* (Gibbon and Ponte, 2005, p77)

At the global level, coordination and control of global-scale production systems can be achieved in different ways. They do not usually take place through direct ownership, though this is the most effective model for more customized products. Multinational companies may find it advantageous to 'outsource' an increasing share of their non-core manufacturing and service activities, and this leads to 'a growing proportion of international trade occurring in components and other intermediate goods' (Gereffi et al, 2005, p80). Lead firms in GCC may then exercise control by developing, disseminating and imposing quality standards and other codification schemes. Dolan and Humphrey (2000), for example, observed such a form of control when analysing the sale of fresh vegetables in the United Kingdom (UK). Large British retailers are organizing and dominating the global supply chains through which fresh vegetables from Africa are traded. These firms exercise control over suppliers by defining quality and packaging conditions and by setting grades and standards. For more elaborate products, more well-developed coordination mechanisms than markets are required. Market-based coordination mechanisms are more effective for standard products and when the costs of switching partners are low. Thus, coordination of global commodity chains may be realized by large multinational companies owning all phases, but more indirect mechanisms through markets, contracts and standards seem to be more attractive.

The GCC approach focuses on the vertical relationships between buyers and suppliers and on the movement of a food item from producer to consumer. Economic relations dominate this type of analysis, while political dimensions are considered to be largely derived from these economic dynamics. When applying the GCC approach in the analysis of current practices in the global provision of food, however, it becomes clear that a coordinative pluralism must be acknowledged, as well as the presence of broader institutional frameworks in which firms operate. Lead firms may shape standards of quality, but 'the broad social norms that stand behind them are, like trade rules, to some extent removed from their influence' (Gibbon and Ponte, 2005, p86). Note that the analysis of global food-supply chains can take place, according to the GCC perspective, independently from the particular countries and institutional frameworks involved. This is why GCC approaches have been criticized for being overly structuralist and dismissive of human agency and the distinctive interests of various actors. Political economy approaches risk obscuring the concrete dynamics between a wide range of political, economic, social, cultural, technological and natural phenomena that extend across localities, regions and nations and that together define globalization (Busch and Juska, 1997; Deeg and Jackson, 2007). Ignoring these particular processes means that GCC approaches have difficulty explaining the diversity present in contemporary food provision at the global level. At the same time, authors such as David Goodman (Goodman and Watts, 1997; Goodman and DuPuis, 2002) and Michael Redclift (1987; Goodman and Redclift, 1991), open up this narrower approach within GCC studies and allow for more flexibility in applying this concept.

## Actor-network analyses of food production and consumption

At least partly in reaction to the emphasis placed on structural characteristics and economic mechanisms characteristic of the political economy view presented above, other, more flexible, conceptual models were introduced, including network-based analyses. One important advantage of these approaches is that the analytical focus no longer concentrates on the producer; the consumer receives equal attention.

The political economy perspective focuses on the study of particular food products, such as tomatoes or beef, within the framework of GCCs. This approach, however, risks reifying the role of corporate actors within these chains, making the roles of other social actors disappear behind the logic of the economic system. Also, nature is usually seen as passive in GCC, but modern agriculture is engaged in a continuous battle on its 'genetic, physiological, biological and ecological limits' (Busch and Juska, 1997, p691), wherein nature is becoming an increasingly active participant as well. One way of opening up this perspective is to apply a network-oriented perspective and to 'focus upon ways in which agriculture and food can be socially embedded in local and regional spaces, as part of local systems and networks' (Marsden, 1997, p189).

Network approaches seem better able to analyse the distributive effects of technological, political and cultural, as well as economic changes (Granovetter, 1985). Building on Latour (2005), such approaches also allow viewing relationships as being mediated by someone or something because institutions cannot operate in a vacuum. Dynamics in food provision therefore must be analysed as specific arrangements of human–human and human–thing relationships. This means we have to 'follow the changes a commodity undergoes as it moves through a (commodity) sub-sector' (Busch and Juska, 1997, p704). A network approach allows for a more detailed analysis of how food is connected both horizontally and vertically in the spaces and networks where the other social and socio-material practices are situated. Social, economic and political dynamics create evolving relationships of dominance and dependency (Marsden, 1997). Network approaches pay attention not only to the vertical coordination within the value chain to optimize the generation of value, but also to the horizontal relations at the same technical stage of production, including social networks, learning and network externalities. This makes it possible to view power as relational and to focus on the (shifting) balances of power and the differentiated access to rules and resources and not to simply look for the dominant powerholder. Hence, varieties of market relationships may emerge, resulting from 'the crystallization of norms and values into rules and institutions, which form the matrix of social relations, be they local, regional, national, or global' (Wilkinson, 2006, p24). As different social levels of interaction influence each other, global processes 'are constantly internalized and layered by different actors in networks of relationships' (Marsden, 1997, p173). All institutions relevant for a particular food product therefore interact with each other and make the subsector work. These points of interaction between institutions are particularly interesting to study, as they 'are the weak links in the chain' (Busch and Juska, 1997, p693).

Markets in general, and especially global markets, lack coordination through formal institutions, and therefore product definitions and requirements among various social actors must be harmonized in other ways. Networks allow for rather complex inter-firm divisions of labour, whereby suppliers produce according to customers' specifications but retain responsibility for their own competencies. According to network analyses, coordination within value chains is 'jointly organized by consumer/ environmental associations, public sector representatives and actors within the relevant filière (i.e. the value chain)' (Wilkinson, 2006, p20). The exact organization of networks is determined by the product characteristics and the processing activities involved.

## Convention theory

Chains and markets may be coordinated through different principles or as the French social scientists Boltanski and Thevenot call them: 'conventions'.

Convention theory, developed by Boltanski and Thevenot (1999),[4] provides an interesting perspective on the coordination and governance arrangements within global food-supply networks. Conventions are based on the fundamental consideration that commercial activities are not possible without a prior understanding among the social actors involved. People require a shared framework of analysis (or a convention) to judge a product's quality, price, etc. before they are willing to consider selling or buying it. The theory focuses on (1) the norms and values that shape divergent assessments of quality, (2) the qualifications, rules and procedures that coordinate exchange relations and (3) the organizational forms that correspond to and uphold particular qualifications. This allows for the combination of different cognitive, normative, inter-subjective and material preoccupations into mechanisms built for managing quality and performance requirements. Food is therefore seen not simply as a material good, but also as an element in wider social, cultural and material practices through which food is produced, processed, traded and consumed. These different practices are linked through conventions. These conventions result from the coordination of situations and the ongoing resolution of differences in interpretation of new or modified contexts of action. Conventions are defined as 'practices, routines, agreements, and their associated informal and institutional forms which bind acts together through mutual expectations' (Salais and Storper, 1992, p174). According to this perspective, trading food is possible because producers know what consumers expect and consumers know what they will get from producers in terms of quality. This agreement need not be uniform for all producers and consumers, while reciprocal expectations may also change over time and between supply chains. There are different and evolving conventions, and there is a direct link between understandings of quality and the social organization of production and exchange. Hence there are multiple conventions and their 'main points of reference are to different types of product, different forms of enterprise and different historical periods' (Daviron and Gibbon, 2002, p144). Diversity in the organization of food provision should therefore be explained not only through differences in technologies, markets and transaction costs, but also through the presence of different conventions, which themselves are not reducible to any discrete and fixed sets of externally given causes. Different (mixtures of) conventions may exist (see Box 2.3) and underpin food-supply networks, and thus various coordinating principles may connect the activities involved in producing and consuming food.

Conventions are continuously negotiated and may even compete, and therefore the balance between different types of conventions within global food-supply networks may change over time.[5]

A network approach to analysing global food provision therefore enables us to acknowledge the importance of quality, as markets can function only on the basis of a prior and shared definition of the quality of the products to be exchanged. Quality definitions are often difficult for consumers to capture simply, and so they are guaranteed by devices, such as brands, labels and social

### Box 2.3 Different quality conventions

Convention theorists have attempted to identify and categorize quality conventions. The following quality conventions related to coordination conventions have been established:

- *Market* coordination, whereby the agreement is based on the economic value of a food product in a competitive market and differences in price are equated with differences in quality.
- *Domestic* coordination, which is based on trust and loyalty, resulting from long-term personal relationships between actors or the use of private brands that publicize the quality reputation of products.
- *Industrial* coordination, which resolves uncertainty about quality through common norms or standards often enforced by an external party via instrument-based testing, inspection and certification.
- *Civic* coordination, which builds on the collective commitment to welfare, implying that all benefit, and whereby the identity of a product is related to its impact on society or the environment. A variation of this is the *green* convention, which occurs when the good of the collective is considered to depend on the general good of the environment.
- *Inspirational* coordination, which consists of conventions based on passion, emotion or creativity that are connected with the notion of a common humanity.
- *Opinion-based* coordination, which is based on public reputation or renown and on the principle of difference.

*Source:* Thévenot (1989)

relations and are consolidated in network arrangements. The presence of such heterogeneous organizational arrangements may persist in the contemporary globalized food supply despite the continuous pressure towards homogenization.

## Combining Political Economy and Network Perspectives

We have examined different conceptual frameworks that can be applied to analyse global food provision, whereby the political economy framework focuses primarily on the distribution of economic power within GCCs, while network perspectives consider the roles of different social actors and how they coordinate their activities within the chain through quality conventions. In this section, we combine both perspectives through the introduction of global production networks (GPNs) and connect this with the concept of flows to better include material dynamics.

The concept of GPNs starts from the premise that networks are key to understanding phenomena such as global food provision, because they reflect 'the fundamental *structural* and *relational* nature of how production, distribution and consumption of goods and services are – indeed always have been – organized' (Coe et al, 2008, p272). This leads Coe et al to define GPNs as 'the globally organized nexus of interconnected functions and operations by firms and non-firm institutions through which goods and services are produced and distributed' (Coe at al, 2004, p471). This constitutes a heuristic device for analysing the highly variable and contingent (both organizationally and geographically) character of the global food supply. GPNs lead to a better understanding of the current dynamics than chain approaches, which consider the global food supply as organized essentially through linear structures from production to consumption. GPNs reveal the more complex and multi-stranded connections among capital, knowledge and people that underlie the production of all goods and services, as well as the various firms that are involved in those circulatory processes. For instance, labour geographies display different dynamics when compared to capital, which is much less place-bounded. By using the term 'production', rather than 'commodity', GPNs prioritize social processes and social interactions (Hughes et al, 2008, p348). This makes room for bringing in supplier firms, service functions, governments, NGOs and social movements, as well as different types of knowledge diffusion when analysing contemporary global food provision, thereby putting the lead firm's role in perspective. Even the firms themselves are not considered to be monolithic, or 'black boxes', but networks themselves 'with varying degrees of imbrications and interconnection' (Coe et al, 2008, p277). In this way, GPNs open the view on the connections, conflicts and synergies among different production networks.

Adopting (global production) networks as the foundational unit of analysis means a focus on relational processes rather than on organizational forms or structures. Material flows connect all network actors as food products move from farmer to consumer while money flows the other way round. Borrowing from commodity culture studies (Cook et al, 1998), GPNs also acknowledge the importance of circuitous knowledge, (re)linking food producers and consumers. Information and knowledge not only flow from producers to consumers, but the other way around as well, encompassing cultural characteristics or food safety and environmental concerns prevalent among consumers. This information and knowledge flow is particularly encouraged by retailers, who 'continually draw upon, transform and circulate product knowledge developed in spheres of both production *and* consumption' (Hughes et al, 2008, p349). Paying attention to this flow makes it possible to understand the cultural politics of knowledge circulation and for a balanced inclusion of consumers and civil-society organizations as co-developers of practices in GPNs.

GPNs are embedded within multi-scalar regulatory systems where national authorities engage with multilateral authorities such as the WTO and with private regulations. So the role of nation states should not be ignored, although international lead firms and quality standards are central in the coordination of GPNs. See Box 2.4 for the growing significance of corporate social responsibility as a form of private governance.

### Box 2.4 Corporate social responsibility

Corporate social responsibility (CSR) entails three main dimensions: profit, people and planet. Companies should do everything within their capacities to promote CSR throughout the business chain of which they are a part. They should take responsibility for the social, ecological and economic consequences of their actions, making themselves accountable and engaging in a dialogue with all those involved.

Consumers seem more willing to purchase products from companies that implement CSR strategies and communicate this through their websites. Among the different industries, the public views the food industry, consumer goods and retailers as performing the best.

*Source:* www.foodnavigator-usa.com/Financial-Industry/Food-industry-well-respected-for-CSR-efforts-Survey (accessed 13 October 2010) and www.mvo-platform.nl (accessed 9 December 2010)

## Sociology of networks and flows to analyse global food

Social scientists have been using the concept of networks to help analyse the process of globalization since the 1980s. In particular, Manuel Castells (1996, 1997, 1998) conceptualizes the transition to global modernity as the evolution of a global network society, where global flows supplant local places as its main characteristic. In this global network society, time and space are structured in new ways because modern transport, information and communication technologies allow activities at distant locations to be connected in a functional unit. Networks may acquire different configurations, through variations in the size and density of their networked connections, as well as in their links with other networks. Networks have a binary character of inclusion and exclusion, and they include only those actors and materials deemed relevant from their internal perspective while ignoring the rest. In contrast to spatial units such as nation states, which include all living within a particular territory, networks only encompass those people and locations that are relevant for their operation.

Global networks are stretched across multiple and distant places and times, so time and place are no longer closely bound (Adam, 2000). According to Castells (1996), we now live in an age of 'timeless time', replacing the biological time familiar from the past and the clock time of the 19th century as the prime rhythmic organizing principle. Under 'timeless time', the conventional sequencing of events is disappearing and familiar rhythms are breaking down. For example, nowadays fresh fruits and vegetables are permanently available in supermarkets while they were only seasonal in the past. Specifics of place are also largely irrelevant, as exchanges and social interactions are increasingly occurring without direct, face-to-face contact between the people involved. Connections between people, machines, texts, objects and technologies interact across multiple and distant times and places and are mediated by various means of communication (Urry, 2003). An illustration is that consumers may nowadays trace back the origin of their food and (virtually) visit the farm through the internet on the basis of information provided on the package of a food product (see Box 2.5).

## Box 2.5 The Coffee Tracer

The Dutch supermarket chain, Albert Heijn, sells its home-branded Perla coffee under the Utz Certified label. Coffee farmers and co-operatives use this Utz Certified label to show that they grow their coffee professionally and with care for their local communities and the environment. Whenever a consumer buys Perla coffee, she or he can trace its origin on the basis of the sell-by date on the package. Through the 'Coffee Tracer' at the company website, the consumer can find extensive background information about the farms or plantations where the coffee was cultivated. Information is provided on the farm and the people working there, as well as on the way social responsibility and respect for the environment are implemented.

*Source:* www.ah.nl/perla/herkomst.jsp and http://consumer.utzcertified.org/index.php?pageID=202&switch language=EN (both accessed 14 September 2010)

In the global network society, we encounter all kinds of configurations of actors, materials, processes and connections. Explaining the dynamics of these configurations relies on images of flows and uncertainty, and dynamics and irreversible change, rather than on older images of order, stability and systemness. Global flows travel at a high speed across national borders, and the emerging global system is not necessarily coherent but often a set of disjunctive dynamics. Food flows offer a concrete example of these complexities (Appadurai, 1996; Urry, 2003). Nevertheless, when flows that connect physically disjointed positions gain some permanence, Castells speaks of the 'space of flows' (Castells, 1996),[6] which may be largely virtual or have more material specificity depending on the content of the movement.

A sociology of global food networks and flows may offer an interesting starting point for analysing dynamics in global food provision (see Spaargaren et al, 2006). Such an approach places networks and flows, rather than localized production systems or GCCs, at the centre of analysis. While this approach addresses the relation between material flows, it also incorporates social institutions and practices, as well as the actors that coordinate and govern them. As such, the focus is not on the characteristics of fixed entities, actors or localities, but on their mutual relationships and the resulting global complexity. The complexity of such transnational flows prevents individual human action from exercising a determining influence on the direction and structure of their movements. Nevertheless, individual agency should not be ignored, so by using a social-practices approach the elements of individual agency and overarching structures can be balanced. Such a structuralist perspective considers individual consumer behaviour in close conjunction with the social and economic structures through which food is provisioned. These social and economic structures are not merely restrictive but enable human agency as well. Social actors are involved in (re)constructing the complex social practice of providing food in multiple ways. Hence, the resulting global dynamics may display contradictory tendencies of which the outcome cannot already be known beforehand.

At the same time, as Castells (2009) argues, the global network society allows unknown forms of power to emerge. Being connected to the network or not is becoming a key (and binary) distinction, because without involvement in a network, people and regions are excluded from the dominant dynamics in contemporary societies. But even among the people that are connected to networks, those performing generic labour should be differentiated from those who 'programme' the networks and the 'switchers' who connect different networks. Generic labour depends on executing tasks defined by others but programmers and switchers have much more autonomy. Programmers combine information and knowledge to define network goals and ways to achieve them, thereby developing standards that determine the rules to be accepted by all involved in the network. Switchers connect different networks by defining common goals and combining resources to strengthen their networked position in society and their operational effectiveness. These new categories of power-holders may be less visible than the economic and political leaders of the past, but they are much more effective in the global network society.

Food is increasingly becoming a global flow that includes transnational and local material dimensions, as well as informational, conceptual and monetary aspects. The nodes in the global food network are connected in a functional and physical sense through the connections of transnational trade. As with global flows of finances and material goods such as computers or cars, food is also travelling all over the globe and processed into products that combine inputs from different places. At the same time, food differs from these other flows, as it exhibits particular tensions between the global space of flows and the local space of places, between the virtual and material dynamics and between the different components of networks. A particular tension concerns the interactions between the flows of food and the natural environment through withdrawals and deposits at different locations (Coe et al, 2008). Food displays a special character compared with other material flows because of its organic character (Morgan et al, 2006). The production of primary agricultural products is limited to specific locations and seasons, while consumers need to bodily incorporate food every day and prefer to do so according to their specific cultural norms and traditions.[7] Therefore, food displays specific local dynamics in the 'space of place' of its production and consumption, and these dynamics include the specific local social organization of both activities. Moreover, global flows of food have particular local, place-bounded impacts as well. This perspective based on the sociology of networks and flows highlights the inherent tension that exists in the context of globalization between dynamics in the space of flows and dynamics in the space of the place of food. See Figure 2.1.

Tensions in global food provision exist between the dynamics in the space of flows and those in the space of place, both at the production end and at the consumption end. Whereas the global space of flows builds on flexibility and links ever-shifting combinations of contributions from distant locations, the spaces of place operate according to specificity and coordination between the present social actors and resources. The tension resulting from these different dynamics is shown, for example, in the pressure on farmers in the global

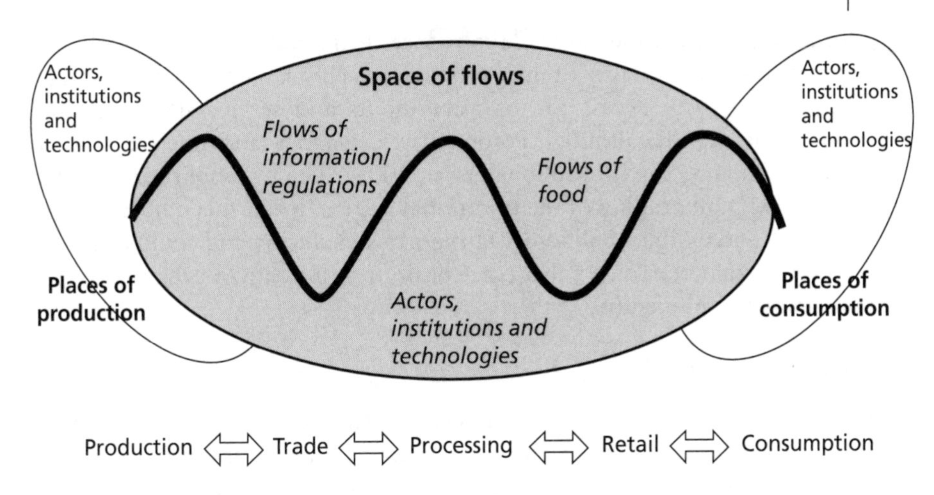


**Figure 2.1** *The 'space of flows' and the 'spaces of place'*

South to adapt their farming practices to fit the particular timeslot that they are assigned in global food provision in the North. Local environmental problems, such as reduced availability of fresh water and over-fertilization resulting from intensive production for the global market constitute another example. On the consumption side of global supply networks, one can point to the increasing availability of standardized food products and the spread of fast-food restaurants (Pollan, 2008), but also to the growth of consumer concerns resulting from reduced trust in contemporary food provision (Kjaernes et al, 2007).

The perspective of global networks and flows on contemporary food provision thus offers useful insights into the familiar and unfamiliar challenges faced nowadays by producers, consumers and government authorities. We will expand here on three important issues: the impacts on small-scale farmers in developing countries, the changing role of governments in regulating food provision, and the persistence of the hunger problem.

## Changing position of farmers through global food provision

Small-scale farmers in developing countries around the globe are confronted with radical changes, including rising incomes, demographic shifts, technological changes in managing food chains and globalization, which they must accommodate. They have to do this while being part of unorganized supply chains with limited market infrastructures and mainly focusing on subsistence agriculture; larger farmers are included in (domestic or export) market-oriented supply chains. Without making transformations, smallholder farmers risk losing out as the second category is much better placed to serve the demands of domestic retailing firms and of exporters selling their produce on the growing global market. The management and composition of production systems is being transformed as retail becomes more organized, wholesale more specialized and procurement more formalized. Smallholder farmers can engage with these transformations only if they are able to reduce their production and marketing costs and improve the quality of their production. This can be achieved through contracts giving them access to necessary inputs and services (Swinnen and Vandemoortele, 2008), although co-operatives, tenancy and nucleus systems could serve similar goals (McCullough et al, 2008). These globalization-related changes allow an upgrading of the traditional system, but if this modernization process is not successful, smallholder farmers risk exclusion and could be forced to remain in production on a subsistence basis or to migrate to urban areas and find other sources of income.

## Governance of food

Food has been subject to government interventions since the 19th century (Tilly, 1975), but the role of governments is also changing under conditions of globalization. Governments nowadays struggle with de-territorialized and decentred mobilities of the numerous material flows in the global network society. De-territorialization is taking place through cultural and economic dynamics, as well as through political and environmental dynamics. States are losing their traditional role as 'a sovereign state, whose hierarchically imposed commands are binding on all parties subject to its jurisdiction' (Karkkainen, 2004, p76). This conventional role is no longer available because any single state nowadays finds diverse self-organizing networks, fluids and 'policies' seeking to striate its internal space and to transform the space beyond it. In the age of globalization, the 'gardening state' (Bauman, 1987), which was chiefly concerned with patterning, regulating, ordering and controlling, is increasingly unable to fully control domestic social processes and can only regulate (the conditions for) mobility. Although sovereignty has not disappeared completely, states must accommodate many other authorities that operate through their own mechanisms. Every state today has to transform into a network state and become part of a complex web of power-sharing and negotiated decision-making processes involving transnational and national political institutions as well as market actors and NGOs. The network state may not need to be formally connected with other states and actors, but it must increasingly coordinate and harmonize its interventions with these others to be effective. In matters of global food regulation, for instance, the involvement of the WTO is growing and public–private partnerships are increasingly present (Mansfield, 2004).

In reaction, critics of this trend promote regulation alternatives that underline locality and specificity in food provision. These alternatives encompass models such as local food-supply systems, where food is sold directly to consumers without interference from intermediaries, at farmers' markets and sales at the farm gate (Renting et al, 2003). The international farmers' movement coalition, La Via Campesina, promotes 'food sovereignty' as an alternative to the present global market-based system of food provision (see Chapter 9). We discuss these mainstream and alternative views on global and local food governance more extensively in Chapter 4.

### Box 2.6 The surge in the price of food

Since the mid-1970s, a sufficient supply of food against ever lower prices seems to be a fact of life for most people in the world, but not for some 900 million people living in Africa and Asia, where the prices of basic food products such as maize, wheat and rice rose dramatically in 2007. Food riots occurred in several cities and agriculture and food returned to the global political agenda and were discussed at global conferences. Rising prices resulted only partly from crop failure, and many considered the growing demand for meat in China and India and the use of crops, such as maize and sugarcane, for the production of biofuels to be the main culprits of this problem. Solving the problem therefore required more than simply increasing production but also an answer to the question of how limited available natural resources, such as land, water and energy, should be used in an optimal and equitable manner. Although agricultural commodity prices fell during 2008, this crisis still made people aware that the era of ever-cheaper food had probably passed, while the political attractiveness of promoting biofuel production was also reduced.

*Source: The Economist* (2007) and Lang (2010)

## Persistence of hunger and the challenge of food security

Today, hunger remains a present threat in several countries (FAO, 2010b), with over 900 million people currently lacking access to sufficient food, and many expect global food security to be a major challenge for the coming decades.[8] Future food security is under threat from a growing global population and increasing affluence for many, citing economic growth in China and India, which translates into an increasing demand for high-value food products of animal origin (dairy and meat). Nowadays, fewer people produce their own food and more depend on markets to secure their food for daily consumption (see Box 2.6).

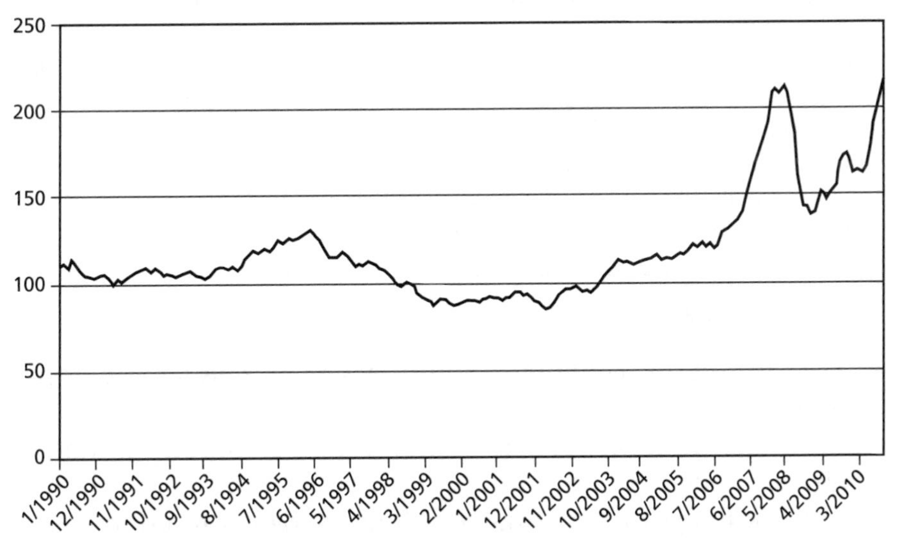


**Figure 2.2** *Trends in global food prices (2002–2004 = 100)*

*Source:* FAO Food Price Index (January 2011)

Sudden jumps in food prices on the global market in both 2008 and 2010 seem to indicate that the secular trend of decline has come to an end. The future will bring at least more volatility on the global market and probably also an upward move (see Figure 2.2).

If these trends continue, hunger may increase and the pressure to increase food production will be more intense than in recent decades. These developments also mean that technological challenges must generate responses in order to increase agricultural production, while the pressure on natural resources, such as land, water, climate, energy and other fossil inputs (such as phosphates), will grow.

## Conclusion

The accelerated globalization of food production, distribution and consumption raises several challenges, including with respect to its sustainability (McMichael, 2000); understanding this acceleration requires the use of adequate conceptual frameworks. In this chapter, we reviewed several potentially helpful conceptual frameworks and concluded that the sociology of global networks and flows offers an interesting point of departure. This perspective can support an analysis of local and global dynamics of food in an integrated manner. By focusing on different networks and their respective external connections, it becomes possible to analyse the involvement of various governmental and non-governmental social actors in the global flow of food. Although much food is still both produced and consumed locally, it is also increasingly directly influenced by global dynamics.

Most conceptual frameworks for analysing global food provision dominated by economic, political or cultural dimensions ignore the organic character of food. Food's material dimension gives it a particular flow, and this should not be ignored when trying to understand problems of sustainability in contemporary globalized food provision. The sociology of networks and flows allows a concurrent analysis of both social and material aspects. One important contribution of this analytical framework is that it makes it possible to identify the fundamental tension between dynamics in emerging global 'spaces of flows' and those in remaining local 'spaces of places'.

---

*Take-home lessons*

- Globalization is changing food provision in several important ways.
- A variety of conceptual tools has been developed to analyse these changes.
- The sociology of networks and flows allows a more in-depth analysis of these changes.

---

## Notes

1 http://comtrade.un.org, accessed 24 January 2011.
2 McMichael (1994, p7) also argues that 'new corporate strategies have emerged, geared to global and regional, rather than national, markets for new specialty products (beyond mass products) oriented to segmented niche markets, such as those for fresh, exotic, and ethnic foods'. This confirms his suggestion for the emergence of a new food regime.
3 GlobalGAP are retailer-based conditions for good agricultural practices. See Chapter 4 for further explanation of GlobalGAP and HACCP.
4 See also Thévenot (2007).
5 Traditional food products, for example, are rooted in domestic mechanisms of coordination and justification and fixed in space (a particular place) and time (a particular tradition). These mechanisms present limits to the expansion of such products.
6 The space of flows still includes a territorial dimension because it requires a technological infrastructure that operates from certain locations connecting functions and people located in specific places. But these network nodes are much less oriented to the specific geographical characteristics of the location and its surroundings and much more to the interaction with other nodes.
7 These norms can be very strict, as, for example, vegetarians do not accept food of animal origin and Muslims do not consume pork (Douglas, 2002).
8 For details on the present global food security situation see: www.fao.org/economic/ess/ess-fs/en

## Further Reading

Bair, J. (ed) (2009) *Frontiers of Commodity Chain Research*, Stanford University Press, Stanford: provides an overview of the evolution of commodity chain theory and research.

Oosterveer, P. (2007) *Global Governance of Food Production and Consumption: Issues and Challenges*, Edward Elgar, Cheltenham and Northampton: elaborates the theoretical model of networks and flows.

Bonanno, A., Busch, L., Friedland, W., Gouveia, L. and Mingione, E. (eds) (1994) *From Columbus to ConAgra: The Globalization of Agriculture and Food*, University Press of Kansas, Lawrence: a classic book on the involvement of multinational corporations in globalizing food.

## References

Adam, B. (2000) 'The temporal gaze: The challenge for social theory in the context of GM food', *British Journal of Sociology*, vol 51, no 1, pp125–142

Anderson, K. (2010) 'Globalization's effect on world agricultural trade, 1960–2050', *Philosophical Transactions of the Royal Society: Biological Sciences*, vol 365, pp3007–3021

Appadurai, A. (1996) *Modernity at Large: Cultural Dimensions of Globalization*, University of Minnesota Press, Minneapolis and London

Bair, J. (ed) (2009) *Frontiers of Commodity Chain Research*, Stanford University Press, Stanford

Bauman, Z. (1987) *Legislators and Interpreters*, Polity Press, Cambridge

Beck, U. (1997) *The Reinvention of Politics: Rethinking Modernity in the Global Order*, Polity Press, Cambridge

Boltanski, L. and Thévenot, L. (1999) 'The sociology of critical capacity', *European Journal of Social Theory*, vol 2, no 3, pp359–377

Bonanno, A., Busch, L., Friedland, W., Gouveia, L. and Mingione, E. (eds) (1994) *From Columbus to ConAgra: The Globalization of Agriculture and Food*, University Press of Kansas, Lawrence

Born, B. and Purcell, M. (2006) 'Avoiding the local trap: Scale and food systems in planning research', *Journal of Planning Education and Research*, vol 26, pp195–207

Busch, L. and Juska, A. (1997) 'Beyond political economy: Actor networks and the globalization of agriculture', *Review of International Political Economy*, vol 4, no 4, pp688–708

Castells, M. (1996) *The Rise of the Network Society. Volume I of The Information Age: Economy, Society and Culture*, Blackwell Publishers, Malden and Oxford

Castells, M. (1997) *The Power of Identity. Volume II of The Information Age: Economy, Society and Culture*, Blackwell Publishers, Malden and Oxford

Castells, M. (1998) *End of Millenium. Volume III of The Information Age: Economy, Society and Culture*, Blackwell Publishers, Malden and Oxford

Castells, M. (2009) *Communication Power*, Oxford University Press, Oxford

Coe, N., Hess, M., Yeung, H., Dicken, P. and Henderson, J. (2004) '"Globalizing" regional development: A global production networks perspective', *Transactions of the Institute of British Geographers*, vol 29, no 4, pp468–484

Coe, N., Dicken, P. and Hess, M. (2008) 'Global production networks: Realizing the potential', *Journal of Economic Geography*, vol 8, pp271–295

Cook, I., Crang, P. and Thorpe, M. (1998) 'Biographies and geographies: Consumer understandings of the origins of foods', *British Food Journal*, vol 100, no 3, pp162–167

Counihan, C. and Esterik, P. (eds) (1997) *Food and Culture: A Reader*, Routledge, New York and London

Dalle Mulle, E. and Ruppanner, V. (2010) *Exploring the Global Food Supply Chain: Markets, Companies, Systems*, 3D, Geneva

Daviron, B. and Gibbon, P. (2002) 'Global commodity chains and African export agriculture', *Journal of Agrarian Change*, vol 2, no 2, pp137–161

Deeg, R. and Jackson, G. (2007) 'Towards a more dynamic theory of capitalist variety', *Socio-Economic Review*, vol 5, no 1, pp149–179

Diop, N. and Jaffee, S. M. (2005) 'Fruits and vegetables: Global trade and competition in fresh and processed product markets', in M. Aksoy and J. Beghin (eds) *Global Agricultural Trade and Developing Countries*, World Bank, Washington, DC, pp237–257

Dolan, C. and Humphrey, J. (2000) 'Governance and trade in fresh vegetables: The impact of UK supermarkets on the African horticulture industry', *Journal of Development Studies*, vol 37, no 2, pp147–176

Douglas, M. (2002) *Purity and Danger: An Analysis of Concept of Pollution and Taboo*, Routledge, Abingdon

*Economist* (2007) 'Food prices: The end of cheap food', *The Economist*, 6 December 2007

FAO (Food and Agriculature Organization of the United Nations) (2010a) *Food Outlook*, November, FAO, Rome

FAO (2010b) *Global Hunger Declining, But Still Unacceptably High. International Hunger Targets Difficult to Reach*, FAO, Rome

Friedland, W. (2005) 'Commodity systems: Forward to comparative analysis', in N. Fold and B. Pritchard (eds) *Cross-continental Food Chains*, Routledge, London and New York, pp25–38

Friedmann, H. (1995) 'Food politics: New dangers, new possibilities', in P. McMichael (ed) *Food and Agrarian Orders in the World-Economy*, Praeger, Westport and London, pp15–33

Friedmann, H. and McMichael, P. (1989) 'Agriculture and the state system: The rise and decline of national agricultures, 1870 to the present', *Sociologia Ruralis*, vol 29, no 2, pp93–117

Gereffi, G., Humphrey, J. and Sturgeon, T. (2005) 'The governance of global value chains', *Review of International Political Economy*, vol 12, no 1, pp78–104

Gibbon, P. and Ponte, S. (2005) *Trading Down: Africa, Value Chains, and the Global Economy*, Temple University Press, Philadelphia

Giddens, A. (1990) *The Consequences of Modernity*, Stanford University Press, Stanford

Goodman, D. and DuPuis, E. (2002) 'Knowing food and growing food: Beyond the production–consumption debate in the sociology of agriculture', *Sociologia Ruralis*, vol 42, no 1, pp5–22

Goodman, D. and Redclift, M. (1991) *Refashioning Nature: Food, Ecology and Nature*, Routledge, London and New York

Goodman, D. and Watts, M. (eds) (1997) *Globalising Food: Agrarian Questions and Global Restructuring*, Routledge, London

Granovetter, M. (1985) 'Economic action and social structure: The problem of embeddedness', *The American Journal of Sociology*, vol 91, no 3, pp481–510

Held, D., McGrew, A., Goldblatt, D. and Perraton, J. (1999) *Global Transformations: Politics, Economics and Culture*, Polity Press, Cambridge

Hughes, A., Wrigley, N. and Buttle, M. (2008), 'Global production networks, ethical campaigning, and the embeddedness of responsible governance', *Journal of Economic Geography*, vol 8, pp345–367.

Ilbery, B. (2001) 'Changing geographies of global food production', in P. Daniels (ed) *Human Geography: Issues for the 21st Century*, Prentice Hall, Harlow, pp253–273

Karkkainen, B. (2004) 'Post-sovereign environmental governance', *Global Environmental Politics*, vol 4, no 1, pp72–96

Kjaernes, U., Harvey, M. and Warde, A. (2007) *Trust in Food: A Comparative and Institutional Analysis*, Palgrave MacMillan, Houndmills

Lang, T. (2010) 'Crisis? What crisis? The normality of the current food crisis', *Journal of Agrarian Change*, vol 10, no 1, pp87–97

Latour, B. (2005) *Reassembling the Social: An Introduction to Actor-Network-Theory*, Oxford University Press, Oxford

Mansfield, B. (2004) 'Organic views of nature: The debate over organic certification for aquatic animals', *Sociologia Ruralis*, vol 44, no 2, pp216–232

Marsden, T. (1997) 'Creating space for food: The distinctiveness of recent agrarian developments', in D. Goodman and M. Watts (eds) *Globalising Food: Agrarian Questions and Global Restructuring*, Routledge, London, pp169–191

Massey, D. (2004) 'Geographies of responsibility', *Geografiska Annaler*, vol 86B, no 1, pp5–18

McCullough, E., Pingali, P. and Stamoulis, K. (eds) (2008) *The Transformation of Agri-Food Systems: Globalization, Supply Chains and Smallholder Farmers*, Earthscan, London

McMichael, P. (1994) 'Introduction: Agro-food system restructuring: Unity in diversity', in P. McMichael (ed) *The Global Restructuring of Agro-Food Systems*, Cornell University Press, Ithaca and London, pp1–17

McMichael, P. (2000) 'The power of food', *Agriculture and Human Values*, vol 17, pp21–33

Mol, A. (2001) *Globalization and Environmental Reform: The Ecological Modernization of the Global Economy*, The MIT Press, Cambridge, MA

Morgan, K., Marsden, T. and Murdoch, J. (2006) *Worlds of Food: Place, Power, and Provenance in the Food Chain*, Oxford University Press, Oxford

Pollan, M. (2008) *In Defence of Food: The Myth of Nutrition and the Pleasures of Eating*, Allen Lane/Penguin, London and New York

RaboBank Nederland (2010) *Sustainability and Security of the Global Food Supply Chain*, Rabobank, Utrecht

Redclift, M. (1987) *Sustainable Development: Exploring the Contradictions*, Methuen Press, London and New York

Renting, H., Marsden, T. and Banks, J. (2003) 'Understanding alternative food networks: Exploring the role of short food supply chains in rural development', *Environment and Planning A*, vol 35, no 3, pp393–411

Ritzer, G. (1996) *The McDonaldization of Society*, Pine Forge Press, New York

Robbins, P., Hintz, J. and Moore, S. (2010) *Environment and Society: A Critical Introduction*, Wiley-Blackwell, Malden and Oxford

Salais, R. and Storper, M. (1992) 'The four "worlds" of contemporary industry', *Cambridge Journal of Economics*, vol 16, no 2, pp169–193

Selwyn, B. (2008) 'Institutions, upgrading and development: Evidence from north east Brazilian export horticulture', *Competition & Change*, vol 12, no 4, pp377–396

Spaargaren, G., Mol, A. and Buttel, F. (eds) (2006) *Governing Environmental Flows: Global Challenges to Social Theory*, The MIT Press, Cambridge, MA, and London

Swinnen, J. and Vandemoortele, T. (2008) 'The political economy of nutrition and health standards in food markets', *Review of Agricultural Economics*, vol 30, no 3, pp460–468

Thévenot, L. (1989) 'Équilibre et rationalité dans un univers complexe', *Revue économique*, vol 40, no 2, pp147–197

Thévenot, L. (2007) 'The plurality of cognitive formats and engagements: Moving between the familiar and the public', *European Journal of Social Theory*, vol 10, no 3, pp409–423

Tilly, C. (1975) 'Food supply and public order in modern Europe', in C. Tilly (ed) *The Formation of National States in Western Europe*, Princeton University Press, Princeton, pp380–455

Urry, J. (2003) *Global Complexity*, Polity Press, Cambridge

Wallerstein, I. (1974) *The Modern World-System; Capitalist Agriculture and the Origins of the European World-Economy in the Sixteenth Century*, Academic

Press, New York

Watson, J. and Caldwell, M. (eds) (2005) *The Cultural Politics of Food and Eating: A Reader*, Blackwell, Malden and Oxford

Wilkinson, J. (2006) 'Network theories and political economy: From attrition to convergence?', in T. Marsden and J. Murdoch (eds) *Between the Local and the Global: Confronting Complexity in the Contemporary Agri-Food Sector*, Elsevier, Oxford, pp11–38

WTO (World Trade Organization) (2010) *International Trade Statistics 2010*, WTO, Geneva

# 3

# Sustainability and Food Production and Consumption

This chapter aims to:

- introduce the concept of *sustainability*;
- discuss key controversies on defining and promoting sustainability in food provision;
- illustrate different strategies and policy instruments to increase sustainability in food provision.

## Introduction

Public concerns about environmental problems emerged and expanded rapidly in the 1960s, and modern agriculture was among the first domains that caused worries. This was not surprising, as producing food had many direct impacts on natural ecosystems and human health. In particular, the use of pesticides and other chemicals in farming drew public attention when the (unanticipated) long-term impact of the extensive use of dichlorodiphenyltrichloroethane (DDT) on birds was discovered by Rachel Carson and presented in her book, *Silent Spring* (1962). Over time, these concerns about the impacts of chemical use in agriculture were supplemented with others, such as animal welfare, food safety, energy use, landscape, climate change and biodiversity. Today, as a result of these issues, food production and consumption have evolved into a multifaceted sustainability concern. Impacts relate to the use of non-renewable (fossil fuels, phosphate) and renewable (solar and wind energy, water) resources and to the impact on the atmosphere and climate (GHG emissions), as well as to soil fertility and land and water management, (agro)biodiversity, pesticide use, waste disposal, economic practices and environmental policies.

In this chapter, we expand on these environmental concerns and situate them in a broader framework of sustainability. We do this by first discussing several perspectives on the relation between food provisioning and the environment. Next, we illustrate the problem by presenting several environmental challenges that modern, industrialized agriculture and food provisioning face. We conclude by discussing several tools that have been developed to increase the sustainability of contemporary food provisioning.

## Modern Agriculture and the Environment: Conceptual Background

Even with widespread agro-industrialization, the agricultural sector remains the largest source of livelihood on the planet, engaging about 1.3 billion people (i.e. roughly one sixth of the world's population). Cultivation and animal husbandry use about one half of all the habitable land on earth (Clay, 2004). As a result of efforts to apply science to increase production, the world's population has doubled since the 1960s, and nowadays for each individual, on average, 25 per cent more calories are available. Households today spend less on their daily food than ever before, in the order of 10–15 per cent of their total net income in Organisation for Economic Co-operation and Development (OECD) countries, while this was still over 40 per cent in the 1950s. 'Even if many developing countries still spend much higher but declining percentages, the diversity, quality and safety of food have improved nearly universally and stand at a historic high' (Fresco, 2009, p379). Every year, additional agricultural land is cultivated worldwide to produce more for a growing and increasingly demanding population.[1] Yet, agricultural producers, whose numbers are declining in many countries, are forced to increase their productivity and aim for economies of scale because food prices have been on a downward trend since the 1970s. This means they are under continuous pressure to use modern technology and find further innovations to grow more food. Increasing productivity is accompanied by a growing demand for fossil fuels and chemical inputs, and the global expansion of agricultural areas is mostly realized at the expense of natural areas, such as forests. A growing number of food producers, consumers and other stakeholders express concern about these serious environmental impacts resulting from the modernization of agriculture, and they call for more attention to sustainability in food provision.

During the 1980s, sustainability surfaced as a policy objective that combined the goals of environmental conservation and poverty reduction. The UN's World Commission on Environment and Development formulated this as 'development that meets the needs of the present without compromising the ability of future generations to meet their own needs' (WCED, 1987, p43). Delegates to the 1992 United Nations Conference on Environment and Development (UNCED) in Rio de Janeiro endorsed this statement, which has remained central in environmental and developmental policies ever since that time. Sustainable development aims to balance ecological, economic and social interests while engaging in efforts to modernize society. Although it is an attractive aim, the exact definition of sustainable development remains elusive and difficult to apply as an overall goal of policy-making (Redclift, 2006). The economy, society and environment are each multi-layered and display fractured dimensions while encompassing different spatial levels as well (Lawrence, 2005). In response to this obscurity, many have tried to introduce a more concrete understanding to make it more applicable in everyday decision-making.[2]

One way of reducing the confusion is to understand sustainable development as a set of guiding principles for environmentally relevant interventions, rather than as an ultimate goal. Carter (2001), for example, characterizes sustainable development as including several core principles in environmental policy: equity, democracy, the precautionary principle, policy integration and planning. *Equity* is relevant, as environmental policy should consider the distributional implications of environmental problems *and* of the policy measures intended to address them. Equity not only concerns poor people in contemporary societies, but also has inter-generational dimensions because of responsibilities for future generations. *Democracy* is important because achieving intra-generational equity requires community participation. The *precautionary principle* guides environmental policy interventions under conditions of uncertainty and argues that a lack of scientific certainty cannot be used as a justification for not preventing environmental degradation. This principle underlines the need to remain on the safe side ('better safe than sorry') when the consequences of technological innovations are not yet sufficiently known. *Policy integration* means that environmental policy cannot be effective as a separate sector but must be integrated in all domains; thus promoting sustainable food provision should be part of not only agricultural policy but also trade and consumer policies. Sustainable development needs *planning* to coordinate different interventions and to manage the complex interdependencies of environmental protection. Such planning needs to take place at different levels of scale because some problems may have a primarily local character while others are more regional or global.

These principles may guide policy interventions, but social actors, who may differ fundamentally on policy aims and environmental values, can also apply them. To better understand these differences, it may be helpful to identify and describe different perspectives on sustainable development.

## Views on Sustainable Development

Over time, many different views on sustainable development have been expressed and for the sake of discussion we group them under a few main headings. These views differ, especially in the ways they characterize the relationship between biophysical environments and humans and the main orientation for intervening. This connection has been extensively discussed in social-science literature. Here we distinguish among the 'neoliberal', 'conservationist' and 'institutional reform' perspectives.

## The neoliberal view

According to the neo-liberal view, environmental problems are best solved through the market mechanism. If natural resources become scarce, their prices will increase and it becomes more attractive to use them efficiently and to invest in finding alternative (resources or technologies). Interference from government will only disturb this process and make it less efficient. Public policies are generally costly and slow in achieving substantial results.

Proponents argue that private companies are best equipped to address agricultural sustainability through their research and development activities. Such research is profit-oriented, and therefore the necessary institutional arrangements need to be in place to make this happen. These companies attach much value to obtaining opportunities to claim patent rights in their discoveries in plant technologies, such as genetically modified organisms (GMOs). Over time, private companies have substantially contributed to increases in efficient food production and processing, but many consider this a one-sided input to the process of sustainable development.

Critics of this view argue that the market ignores not only the interests of those without adequate buying power, such as the poor, but also natural resources like biodiversity (Polanyi, 1944). If the market only were to direct the process of sustainability, it would only profit those parties that already have a strong position. Moreover, the neoliberal perspective tends to neglect the importance of institutions to make markets function (see Chapter 2).

## The conservationist approach

The conservationist approach puts ecosystems first and argues that the natural environment sets clear limits to human activities. Thus the fundamental problem of contemporary society is that these limits are not respected and that economic and technological optimization prevail (see Box 3.1).

To address sustainability issues from this perspective, scientists should clearly determine the limits the environment poses to human behaviour so that governments can confidently impose measures to prevent crossing them. Environmental movement organizations should then carefully monitor this process and pressure governments whenever they fail to fulfil their responsi-

### Box 3.1 The dangers of ignoring environmental limits in agriculture

In a column in the *New York Times* on 16 December 2007, Michael Pollan argues that when people try to rearrange natural systems along the lines of a machine or a factory, whether by raising too many pigs or too many almond trees in one place, whatever they may gain in industrial efficiency they sacrifice in biological resilience. Hence, the question is not whether systems this brittle will break down, but when and how, and whether when they do, people are prepared to treat the whole idea of sustainability as something more than a nice word.

*Source:* Pollan (2007)

## Box 3.2 Thomas Malthus

Thomas Malthus (1766–1834) is regarded by some as the first environmentalist because he argued that the natural limits to producing food would force humanity to adapt their behaviour. In his 'Essay on the Principle of Population' (1798), he pointed out that while the human population increases exponentially, food production can only grow on a linear basis; hence it inevitably leads to catastrophe. Wars, famine, destitution and disease will occur and reduce population numbers. To prevent such tragedies, humanity should exercise more self-constraint.

Neo-Malthusians, who apply Malthus' argument in a more sophisticated way, argue that continuing population growth threatens to deplete natural resources and is hence the most fundamental cause of the environmental crisis.

*Source:* Robbins et al (2010)

bility. Some, called neo-Malthusians, base their thinking on the ideas of the 19th-century thinker Thomas Malthus (see Box 3.2).

Since the 1960s, several others have revised, refined and adapted this conservationist perspective. For instance, in a 1968 essay, Garrett Hardin wrote about the 'tragedy of the commons', warning that without restrictions imposed from the outside, human beings would strive for overusing unregulated commons (i.e. natural resources) because the benefits are individual while the costs for overuse are shared by the community as a whole.[3] This perspective was also applied in 1972 by the Club of Rome, a group of scientists who wrote the influential report 'Limits to Growth' (Meadows et al, 1972), in which they pointed to the limited availability of natural resources, which should force humanity to restrict their use for economic development. In reaction, some argue that the only way forward is to adapt present lifestyles to the limits of nature and the environment set for mankind. They call for a process of 'de-modernization', whereby societies rely much more on small-scale, local communities and human labour: 'small is beautiful' (Schumacher, 1973).

In the domain of food provision, this conservationist perspective has contributed to such strategies as integrated pest management, low external input agriculture and reduced water use. At the same time, this view is rather inflexible, and by 'naturalizing' the limits, it fails to take into account the adaptability of societies in making use of the elasticity of production based on human ingenuity and technological innovation.[4]

The relationships between human society and the natural environment are complex, and thus knowledge about nature does not automatically prescribe the way society should be organized. For instance, as Boserup (1985) shows, technology and its further development through modernization can have positive impacts on the environment, as well as negative ones.

## The institutional approach

Another perspective on sustainability is the institutional approach. This perspective agrees with the ecosystem-first view that contemporary societies are crossing the limitations of nature. Its proponents argue, however, that this results from the ways in which institutions function in today's societies. This view brings together several points of view, including Marxist perspectives and those of scholars arguing for ecological modernization.

Marxists see environmental problems as resulting from the capitalist character of the present economic system, where private profits take precedence over the protection of public goods. They argue that the mechanism of the free market, where natural resources have no price, does not provide a good tool for protecting the environment. Referring to the work of Marx in the 19th century, they point to the presence of a '*metabolic rift*'[5] that is the ultimate cause of environmental problems.

Scholars who support this view insist on the need for institutional change to better secure the environment. This includes exerting more control over large multinational corporations in the global food supply (Roberts and Grimes, 2003; Roberts and Hite, 2007) and addressing the imbalance of power among different groups in society, making some groups suffer more from environmental problems than others (see 'political ecology' in Greenberg and Park, 1994).

Some consider this perspective too rigid and call for a more flexible approach, such as that proposed by 'ecological modernization'. Proponents of ecological modernization argue that concrete social practices, socio-technological systems, institutions and governance arrangements should be studied when trying to understand environmental problems and identify potential responses. According to this view, present society can be made more environmentally friendly by incorporating ecological elements into different relevant social practices. With the help of science and technology, through integrated approaches, environmental considerations could be incorporated into the design, production, consumption and final disposal of all products and services, including in agriculture (Tilman et al, 2002). Markets can play an important role in transmitting ecological ideas and practices, with producers, intermediaries and consumers all playing their parts. The government can encourage this internalization through the use of market-based instruments, such as eco-taxes and tradable permits (Mol et al, 2009).

Critics of this ecological modernization perspective claim that it does not sufficiently acknowledge the presence of natural limits in the environment and that it would be rather naïve to expect solutions from technological change, because it is what created environmental problems in the first place.

## Conclusion

Definitions of sustainable development have generated considerable confusion. In this section we introduced three, still quite general, views. These perspectives are applied by different social actors, and although they differ in some important respects, they all address current environmental problems, including those in agriculture and food provision. Next we discuss several problems related to sustainability in modern agriculture.

## Modern Agriculture and Sustainability: Several Problems

Modern food provision has important consequences for all three dimensions of sustainability: social, economic and ecological. The social dimension is evident, as food security, defined as having access to sufficient and adequate food, remains a challenge for hundreds of millions, if not billions, around the world (Helms, 2004). At the same time, obesity is a growing problem in many countries. In economic terms, agriculture is the main economic activity for millions of households and an important driver for economic growth in developing countries. The ecological dimension of modern agriculture is visible in its direct and indirect interaction with nature, using large quantities of energy, water, minerals, fertilizer and pesticides, and with substantial impacts on natural resources, climate, biodiversity, animal welfare and rural landscapes (Pretty et al, 2000; Aiking and Boer, 2004).

Concerns about the environmental, economic and social impacts of food production and consumption attract much attention in society. Although the exact content of these concerns vary, they have generated (sometimes intensive) public debates about the way food is produced, leading to proposals for radical reforms in food policies and practices. Others prefer less radical change and argue for modifications to reduce the negative social and environmental impacts of present practices (Johnson, 2006). Which way forward is chosen depends, at least in part, on which problem is considered most important. Over the years, evolving sustainability issues have raised public concerns and repeatedly resulted in changing food production and consumption practices. An example of the multiple environmental impacts related to producing food from aquaculture can be found in Box 3.3.

Some environmental concerns are directly related to agro-industrial modes of primary food production, such as soil erosion; the intensive use of water, pesticides and fertilizers; the emissions of different pollutants into the natural environment; the declining attractiveness of the landscape; reduction of biodiversity; and unethical treatment of farm animals (Kirchmann and Thorvaldsson, 2000; Pretty et al, 2000; Horrigan et al, 2002). Others have more to do with the way food is processed and transported and relate in particular to the use of energy and chemical inputs. Food safety should not be ignored either, because responses to secure it may have direct and indirect environmental consequences.

### Primary food production

As the primary stage in the production of food, agriculture has many environmental and social impacts. Agriculture makes use of land, water and other natural inputs, and during human history, technological innovations have made it possible to increase its productivity.

Still, new land is cleared for agriculture, resulting in the reduction of land available for natural ecosystems. Forests in Brazil, Indonesia and other countries

## Box 3.3 Environmental impacts from aquaculture

Modern forms of intensive aquaculture are repeatedly criticized for the environmental damages they create. These include the impact of shrimp farming on mangrove forests and other coastal zones and the use of captured fish to feed farmed fish, such as salmon. For instance, it is estimated that an average of 2–5kg of wild fish is necessary to raise just 1kg of farmed species. This so-called 'fish-trap' means that aquaculture ultimately runs into its own limits and therefore cannot be considered a sustainable solution. An alternative to the use of original fishmeal through adding proteins of plant and animal origin to smaller quantities of processed fish inputs remains problematic because of the limited availability of the correct proteins and palatability problems.

In addition, critics point to the water pollution that occurs through the discharge of water from ponds and net pens and the introduction of exotic fish species, which threatens biodiversity. The local environmental impacts of aquaculture correlate with the intensity and scale of production, in particular, when large numbers are concentrated in protected areas with insufficient water exchange. Fish farmers also must maintain a minimum water quality level, however, to keep their activity productive and prevent massive damage.

*Source:* Pillay (1992)

are cut to find more land for farming. Forests are important for maintaining the global climate, while biodiversity is threatened by deforestation.[6]

Producing food requires the use of inputs, including seeds, fertilizer, machines and pesticides. The actual quantities needed vary widely among different farming systems, while some do not use certain inputs at all, such as organic farming, where (inorganic) chemical inputs are not allowed. Nevertheless, several problems related to agriculture can be identified.

### *Water*

Agricultural production and processing are not possible without water, but the intensification of agriculture also results in significant increases in its use. The intensive use of water in agriculture has led to several environmental problems. Levels of groundwater, lakes and rivers may diminish because of water abstraction; salinization occurs as the result of intensive irrigation; wetlands may disappear; watercourses may be diverted for irrigation; and water use for spreading fertilizers and pesticides may lead to groundwater pollution through leaching (Strosser et al, 1999).

Water (in combination with wind) may also contribute to the erosion of agricultural soils, which leads to the disappearance of topsoil. Innovative methods such as no-tillage farming, whereby the topsoil is disturbed as little as possible, are introduced to reduce this risk.

### *Pesticides*

Pesticides attract much attention because they (may) have direct impacts on human health and ecosystems. The *Toronto Globe and Mail* reported on 15 October 2002 that 'about 20 per cent of the food we eat is contaminated with trace amounts of pesticides, even though most of them have been banned for decades ... such as DDT and dieldrin' (Picard, 2002, pA10). Pesticide use has spread throughout the world, and 1.5 million tons of pesticides are produced every year (Eddleston et al, 2002). Pesticides were adopted widely in the 1950s because they allowed rapid yield increases, but after some decades the negative impacts became apparent. The many health and environmental costs of pesticides could no longer be ignored, while the increasing pest resistance that resulted in lower yields compounded the pressure to identify alternative means of plant protection (Eddleston et al, 2002). Two strategies resulted from this. First, since the 1980s, alternative technologies have been sought to reduce the use of dangerous pesticides and rely more on more selective means, while increasing the application of integrated pest management where pesticides are used much more selectively. Second, farmers who decided to follow organic farming principles have been refraining from all (chemical) pesticides. Still dangerous pesticides remain in use (sometimes illegally) because of their low price and seeming effectiveness, particularly in countries with weak regulation in this domain (Hoi, 2010).

### *Fertilizers*

Crops require fertilizers to secure yields, and they can be of organic or chemical origin. Chemical fertilizers, in particular, may have negative environmental impacts because of the substantial quantities of energy needed to produce them and because of the effects of using increasingly rare natural inputs, such as phosphorus.

Phosphorus is a non-renewable resource that is being mined at an increasing rate to meet the growing demand for artificial fertilizers. In all, chemical fertilizers account for 80 per cent of phosphates used globally (Ecosanres, 2008). Estimates of the remaining amount of phosphorus vary, and projections about how long it will take to deplete this irreplaceable resource entirely range from 50 to 130 years. Therefore, phosphorus use needs to be reduced and recycled where possible, because as it is essential in the plant's growth cycle, agriculture cannot do without it.

Another environmental problem resulting from the use of chemical fertilizer is caused by the potential movement of unused or excess fertilizer through the soil profile into groundwater (leaching). The actual risk of leaching depends on the type of soil, with movement being greater on sandy as compared to clay soils.

### *Energy*

The mechanization that accompanied modernization in agriculture has resulted in huge increases in labour productivity because manual labour has been replaced by machines. For instance, instead of the 1200 hours of labour needed in the US to produce 1 hectare of maize in the past, nowadays only 11 hours is needed (Pimentel, 2009). This process has occurred rather unevenly across the globe, however (see Table 3.1)

**Table 3.1** *Indicators of agricultural mechanization (2003)*

| *Designation* | *Africa* | *Latin America* | *Europe* | *World* |
|---|---|---|---|---|
| Tractors | 537,917 | 1,765,242 | 10,737,469 | 25,530,184 |
| Agricultural workers per tractor | 394 | 24 | 3 | 51 |
| Hectares per tractor | 2113 | 67 | 45 | 187 |

*Source:* Mrema et al (2008)

Nevertheless, the extensive use of machines in farming places a substantial claim on energy, and despite the continuous increase in energy efficiency, the total amount of energy required for agriculture is still growing. More machines are used and much energy is also needed to produce artificial fertilizers and pesticides.

This intensive energy use makes agriculture an important contributor to global warming (see Chapter 5), and it also makes agriculture vulnerable to fuel price increases and to the effects of dwindling fuel resources.

### *Monocultures*

Industrial agriculture has made intensive use of modern technologies to achieve higher productivity, but this intensive farming has brought monocultures (i.e. producing one single crop over a large area) and a declining diversity in landscapes. Consequences of monocultures include their fragility against diseases and their contribution to the overall reduction of *agro-biodiversity*, i.e. the genetic diversity within a particular crop. Reduced variation in landscapes makes the countryside less attractive for both inhabitants and tourists, even though this might generate a useful source of income for farmers. Utilizing 'high-input' technologies, modern agriculture requires more land and fewer workers, resulting in displacement of traditional, subsistence agriculture, massive out-migration from agricultural areas, rapid urbanization, growing hunger and food insecurity. This process is closely related to the agricultural modernization process called 'the Green Revolution' (see Box 3.4).

## Box 3.4 Green Revolution

The Green Revolution was an international campaign supported by the Ford Foundation and others in the 1960s aimed at increasing the productivity of agriculture by means of inserting science-based technologies into the production of food grains, such as rice, maize and wheat. Its objective was to eradicate hunger by introducing innovative technologies and facilities in traditional agricultural societies. The Green Revolution is acclaimed for its essential contribution towards increasing the world's food production and criticized for its negative environmental and social impacts in some parts of the world. In particular, small farmers seemed to have trouble accessing the complete package of fertilizer, pesticides, machines, water and labour necessary to cultivate the new high-yielding seeds.

*Source:* Atkins and Bowler (2001)

### *Animal welfare*

Animal welfare is an issue that has been attracting much more attention in recent years. The treatment of farm animals, such as the procurement of massive farm feeding lots (known as confined animal feeding operations, or CAFOs) and intensive poultry raising, is criticized for not respecting the rights of animals and for endangering human health. Animal rights are nowadays recognized by many and, for instance, the EU officially holds that animals should not be harmed unnecessarily. Intensive animal rearing may result in spreading diseases that endanger human health. In reaction, the chemical substances that are used to prevent their spreading may again threaten health, for instance the intensive use of antibiotics risks the speedier resistance of bacteria.

## Conclusion

The environmental impact of food production should not be approached in a simplistic way, as, for example the International Food Policy Research Institute (IFPRI) observes that:

> *it is commonly thought that intensification of agricultural production usually leads to environmental degradation. [However], in most developing countries too little intensification is a major cause of natural resource degradation, as desperately poor farmers mine soil fertility and climb the hillsides in an effort to survive ... Agricultural development, poverty reduction, and environmental sustainability are likely to go hand in hand when agricultural development is broad-based, market-driven, participatory and decentralized, and driven by appropriate technological change that enhances productivity.* (IFPRI, 2002, p23)

Sustainable food systems can be achieved if more food is produced with less waste and in a manner that takes account of the new nutritional picture dominated by the coincidence of over-consumption, under-consumption and mal-consumption (Lang, 2010). Reliance on single-technology solutions is unlikely to realize these aims and to resolve the complex array of problems ahead, which are partly social, partly environmental and partly related to who controls food systems.

## Food processing

Food-processing industries may also lead to critical environmental problems, such as air and water pollution, the production of large quantities of solid (often organic) waste and considerable amounts of energy used in these industries.

Barrett et al (2001, p423) observe that 'vertical co-ordination through contracting or organizational integration permits downstream interests to exert unprecedented influence over farming practices in which they are not directly engaged'. This influence may have negative environmental effects, for example,

## Box 3.5 Unilever

In 1997, Unilever, a UK-Netherlands-based food company, initiated a programme to apply sustainability principles in its international agricultural supply chains. The company chose to adopt the following goal for this Sustainable Agriculture Program:

> *Unilever will buy all its agricultural raw materials from sources applying sustainable agricultural practices, so that*
>
> - *Nature and biodiversity are protected and enhanced*
> - *Soil fertility of agricultural land is maintained and improved*
> - *Farmers and farm workers can obtain a liveable income and improve living conditions*
> - *Nitrogen fertilizers are used efficiently and don't harm the environment*
> - *Water availability and quality are protected and enhanced*
> - *Greenhouse gas emissions are reduced.*

According to a scientific assessment of the results after a decade, the strategy was effective in some cases, particularly regarding the use of pesticides and the assessment of associated risks. In other cases, there was need to better identify the most critical areas for suppliers, because otherwise farmers would be forced to improve their environmental performance to the detriment of their social and economic conditions.

*Source:* Pretty et al (2008a, 2008b) and Unilever (2010)

when aesthetic requirements stimulate increases in pesticide use. At the same time, however, vertical coordination in agri-food chains may induce environmentally friendly practices in order to achieve better-quality foods that will yield premium prices. The net effect of this influence from agro-industries remains an issue of empirical verification (see Box 3.5).

Next, the food industry may take steps to improve its environmental performance in the production process by reducing energy and water use. This may result in measures that consumers do not expect, such as food processing's sometimes relatively small contribution to the final environmental burden of a food product. Communicating this to consumers is difficult, and therefore companies often choose to keep such conclusions internal.

Larger corporations seem to be more inclined than smaller firms to introduce voluntary compliance to official and private sustainability standards because they are more susceptible to official monitoring and to public scrutiny with the potential to endanger their brand image.

### Distribution and consumption

Distribution and consumption practices may also have significant environmental and social consequences. Wastes are produced and energy utilized for food transported over long distances (so-called 'food miles'). Large retailing firms are often held responsible for these environmental problems because of their decisions regarding the packaging of food in supermarkets and transporting food products from all over the globe. Nevertheless, they can also become a potential leverage for changes in agri-food chains. Konefal et al (2005) observe contradictory tendencies in the food retail sector, where on the one hand, concentration tendencies lead to a global oligopoly in the sector, while on the other hand, increased consumer pressure results in pressure on food retailers to incorporate social and ecological attributes into their production practices (Oosterveer, 2007; see also Chapters 6 and 11 of this book).

Much hunger in the world today is a result not of the quantity of agricultural goods produced, but rather of their inequitable distribution. The world's largest agricultural producers, including the US and some European countries, sometimes literally flood world markets with certain agricultural commodities at very low prices (i.e. 'dumping'). Local farmers cannot compete under such conditions, and they may be forced to leave agriculture. Driven out of the countryside and away from subsistence agriculture, the world's poor often cannot afford to buy even cheap, commercially produced food products. Food aid often does not solve the problem either, as despotic governments may horde food supplies, or distribute them first to armed forces and government employees.

The choice of diet also has far-reaching, interconnected ramifications, both for the environment and for human health (Duchin, 2005). A transition in affluent societies from an American- to a Mediterranean-type diet would have favourable impacts on health and the environment.

## Food safety

The global food supply is often considered responsible for threats to food safety. For example, growth in international trade has meant that pathogens that were once confined to a particular geographical region can now travel around the world on aeroplanes in a matter of hours. In addition, mistakes in the production or processing stages of food, such as adding melamine to milk in China, can have wide-ranging consequences (Xin and Stone, 2008). The increasingly concentrated nature of contemporary agri-food production and distribution exposes far greater numbers of people over much wider geographical areas to contaminated products than in the past. Leslie and Reimer, for example, argue that:

> *the pollution of ecosystems, the growth of genetic engineering of food products and the absorption of chemicals into the bodies of producers and consumers of food mean that there are ethical connectivities between actors at one location in the chain and those at other sites.* (Leslie and Reimer, 1999, p408)

For instance, the discovery of the first case of bovine spongiform encephalopathy (BSE) in the US in 2003, and the subsequent trade restrictions by Japan, resulted in an immediate 18 per cent decline of beef exports from the US to Japan (Leuck et al, 2004). Whether the food-safety risks resulting from the globalization in food provisioning actually lead to more casualties remains extremely hard to determine because it proves difficult to establish the exact number of victims from contaminated food (Nestle, 2002).

Food-safety issues have a serious bearing on consumer behaviour and thereby inevitably also on governance arrangements dealing with these matters. Governments come under pressure to introduce more effective forms of food-safety governance. They respond by trying to achieve better control over (potentially) contagious animal diseases and food risks under globalized food production and consumption and promote rapid information exchanges and strengthened international coordination (Maxwell and Slater, 2003). Despite these attempts to increase governmental control over the food supply, the growing number of food-safety incidents and scandals seems to only incite public concern, which leads to discussions about what regulatory policy changes are needed. Larger industrial food processors respond and may use food safety as a tool to increase their market share (Busch, 2000). This global competition between large industrial food processors and food retailers may reduce the scope for different national food-safety politics.

## Measuring Environmental Impacts: Life-Cycle Analysis

Environmental sustainability is a difficult concept to grasp, and therefore measuring it is challenging as well. Nevertheless, various measurement tools, such as life-cycle analysis (LCA), ecological footprint, food miles, water footprint and $CO_2$ measurement have been developed. The first three of these are explained below and the last one in Chapter 5.[7] See Box 3.6 for an example of food miles.

### Life-cycle analysis

Among the different measurement tools, LCA is considered to be the most complete to measure environmental performance. In essence, an LCA models

> **Box 3.6 Food miles and organic food**
>
> In 2007, the Soil Association, the UK's organic food certifier, was considering a ban on air-freighting of organic fruit and vegetables to reduce carbon emissions from organic produce.
>
> This intention sparked a heated debate, in the UK and beyond, on the impacts such a decision would have on organic farmers in developing countries. Local communities in the world's poorest countries would lose a lucrative market and suffer from a large drop in income. 79 per cent of organic produce from overseas is said to originate from poor countries, such as Egypt, Kenya, Ghana, Zambia and Morocco. At the same time, air-freight imports account for some 3.1 per cent of all organic fresh fruit sales and 13.9 per cent of organic fresh vegetables, or 8.1 per cent of all organic fresh produce sales.
>
> *Source:* www.foodnavigator.com Europe (accessed 5 October 2007)

## Box 3.7 Carbon reduction label

Consumers in the UK can judge the carbon footprint of many goods available in the shops by checking the carbon reduction label. This voluntary, private label was developed in 2007 by the Carbon Trust in partnership with DEFRA (the Department for Environment, Food and Rural Affairs) and the British Standards Institution (BSI). Companies can use the label on a specific product when they calculate its exact footprint to the PAS 2050 standard (see Box 5.4). This calculation offers a more objective basis for a comparison between different products than simply the distance a product has travelled. Once the carbon footprint of a product has been measured and certified, the company then has to commit to reducing the product's emissions. Every two years, the product must be reassessed and a reduction has to have been achieved and independently certified – or the label is removed.

By early 2011, one of the companies involved, Tesco supermarket, had 120 products with the carbon reduction label available in the shops, including potatoes, orange juice and milk.

*Source:* http://business.carbon-label/business/measurement.htm and www.tesco.com/greenerliving/greener_tesco/what_tesco_is_doing/carbon_labelling.page (accessed 31 March 2011)

the environmental impact of a particular product from start to finish. With the help of an LCA one may, for instance, measure climate change arising from GHG emissions, eutrophication as a result of nutrifying emissions (such as nitrate, ammoniacal nitrogen and phosphates), and abiotic and biotic resource depletion (Foster et al, 2006). This measuring is based on quantifiable flows, which makes the results clear and quantitative. See Box 3.7 for an example of a carbon label.

According to the official guidelines, standardized in the ISO 14040 (1997) series, four steps need to be differentiated in an LCA:

1. defining the goal and scope of the LCA (choosing the functional unit for the analysis);
2. conducting a life-cycle inventory analysis (bringing together available data about the environmental impact throughout the life-cycle of a product);
3. creating an impact assessment (scoring the environmental effects grouped in different categories);
4. interpreting results to identify the most important environmental impacts and define concrete action.

The LCA method was initially developed for use on industrial products, so applying it to food offers several challenges. The selection of indicators constitutes the first problem, as many different ones could be used (Sonesson et al, 2005; Foster et al, 2006), such as energy use, global warming potential, acidification, photochemical ozone formation, eutrophication, and total material and resource use. Next, to be applicable, the units of analysis should be appropriate to everyday consumer behaviour, which requires multi-product activities throughout and implies choices about the allocation of flows, while such choices strongly influence the outcomes of an LCA. The main elements and steps in an LCA when applied to measuring the environmental impact of food provision are illustrated in Figure 3.1.

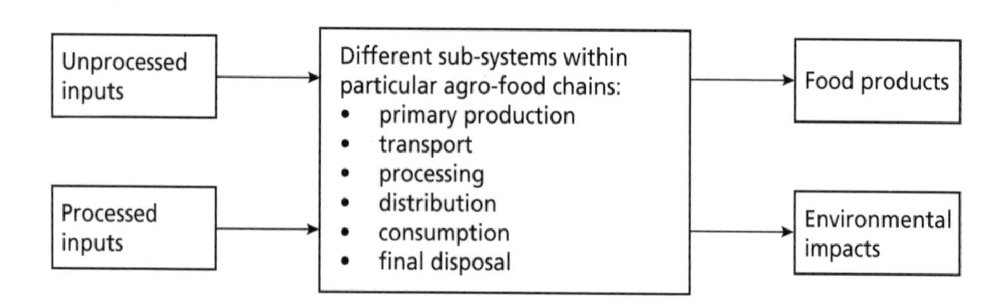


**Figure 3.1** *Model of an LCA in a food-supply chain*

Bread can be used to illustrate this model. Bread has important environmental impacts in its production stage (cultivating wheat), in the processing stage (baking) and in the transport involved at different stages in the supply chain. In the cultivation phase, significant eutrophication impacts may occur, which is linked to leakage of nitrogen from fields, global warming and emissions of nitrogenous compounds (in the production of nitrogen fertilizers and the use of tractors). In the processing stage, baking, environmental impacts are related to energy use, photo-oxidant formation (through ethanol release) and global warming. In the transport phase, global warming and acidification impacts are highly relevant. When these impacts are carefully measured with the help of an LCA, comparisons can be made between different technological options in the production process (Foster et al, 2006) (see Box 3.8 for an illustration).

LCA-based comparisons may be very instructive, as they sometimes undermine established heuristics that are applied to differentiate between sustainable and non-sustainable food (see Box 3.9 for an illustration).

In practice, however, very few fully elaborated LCAs measuring food's complete environmental impact from 'farm-to-fork' exist because of the complications involved. The selected baseline used for comparison is often arbitrary. Determining the system's boundaries is also particularly difficult, because food provisioning today has many indirect impacts that may be highly relevant. For instance, it is not clear whether the environmental impacts of producing farming machines should be included in an LCA or not, or whether the consumption stage should be part of it and how this could be standardized. Foster et al (2006) observe that most studies on the environmental impacts of food production and consumption in the UK focus on the farm end of the production chain and pay much less attention to what happens beyond the farm gate. An LCA becomes even more complicated when it concerns calculating the different impacts in a composed product, such as a pizza. It is not clear how the units for calculation in the different measurements should be established and how to avoid the one-sidedness of a single indicator. Another challenge is how to add up the scores in different dimensions (Kessler et al, 2007) (see Box 3.10). Some try to use money or $CO_2$ emissions as a universal standard, but some impacts are hard to calculate in such terms and doing so demands many different assumptions.

### Box 3.8 Applying an integrated environmental analysis

DEFRA in the UK commissioned a project to develop a model for the mass and energy flows arising from the production of ten key commodities and to identify high-risk parts of the system. The resulting model should make it possible to compare resource use and emissions arising from various production options in England and Wales. All inputs into on-farm production for each commodity are traced back to their primary resources, such as coal, crude oil and mined ore. All activities supporting farm production, such as feed production and processing, machinery and fertilizer manufacture, fertility building and cover crops, are included. Soil is included to a nominal depth of 0.3m. When appropriate (for example, for tomatoes and potatoes), commodities are defined as *national baskets of products* and are included in their proportion of national production. Abiotic resources used are consolidated into one scale based on relative scarcity. Individual emissions are quantified and aggregated into such impacts as global warming, eutrophication and acidification. Organic production systems are analysed for each commodity, as well as variations on non-organic (or contemporary conventional) production. Interactions among inputs, outputs and emissions are represented by functional relationships derived from process models wherever possible, so that as systems are modified, they respond holistically to specific changes.

This model is filled with data, although finding them is often complicated because they do not exist or are commercially confidential. Still the authors claim that, despite a margin of uncertainty, the model is very useful for comparison purposes. They conclude that organic field crops and animal products mostly consume less primary energy than their non-organic counterparts because they use legumes that fix nitrogen rather than synthetic fertilizers made from fossil fuels. Poultry meat and eggs are exceptions, because the overall efficiency of feed conversion in the non-organic sector is higher. The relative burdens of other environmental impacts, such as global warming, acidification and eutrophication between organic and non-organic field-based commodities are more complex. Organic production often results in increased burdens, from factors such as nitrogen leaching and lower yields. The lower yields and fertility building of organic production mean that between 65 and 200 per cent more land is required for organic production.

*Source:* Williams et al (2006)

A less simplistic and more comprehensive approach than an LCA is needed to assess the sustainability impacts of the global food supply. Such a measurement should include the three dimensions of sustainable development – social, economic and environmental – and it also should recognize the existence of comparative advantages and specialization. In the end, such an approach would not result in one single indicator, but a matrix where the scores on different dimensions are presented in a simplified manner.

## Ecological Footprint

A rather familiar environmental indicator is the ecological footprint, which has become an important and useful communication tool (Wackernagel and Rees, 1996; Rees, 2003). Ecological footprints indicate the quantity of land needed to provide resources for a specific activity, for example, for food consumption (Gerbens-Leenes et al, 2002). They are aggregate indicators of the demand on

### Box 3.9 'Quinoa, organic but not sustainable'

*Quinoa is a small grain produced in Bolivia. Because quinoa has become very popular recently, its export has tripled over the last three years. Shifting from a niche product sold via specialized shops to large sales through supermarkets has led to a serious problem: the intensification of production. Even those that qualify as organic result from monocultures that have invaded the Bolivian plains at the expense of herding llamas and growing food crops for subsistence.*

*The loss of soil fertility, due to a substantial increase in areas devoted to the same plant, do not guarantee the sustainability of this crop, not even if it is organic.*

*Returning to a balanced agricultural system would allow producers to diversify their income and secure their future. But market demands do not pay attention to the long-term impacts on such culture.*

*Therefore, even when you are seduced by the enjoyable qualities of this grain, limit your consumption, to reduce your ecological impact and for the welfare of the Andes.*

*For this reason, we have stopped selling several products based on quinoa. Thank you for your understanding.*

*Source:* A leaflet handed out at a farmers' market in the south of France, summer 2009 (translation by the authors)

nature expressed in 'global hectares' (gha), usually per capita (gha/cap). Household food consumption is assessed based on household expenditures and prices of particular consumption items. Not only land but also energy requirements are combined for these items. The ecological footprint of food at the household level is calculated by combining household consumption (kg) with land requirements per food item ($m^2$/year/kg). This requirement can be compared with the mean total area available for all of the world's population (1.8gha/cap). The calculation can also be done individually, for a region or for a country. As typically utilized, what ecological footprints do not do so well is bring attention to the disproportionate consumption or 'footprints' within a given population,

### Box 3.10 Using LCA methodology on global food commodities

Kessler et al (2007) applied the LCA methodology for a comprehensive assessment of the sustainability impact from global commodities. In a first attempt, they addressed biodiversity as well as the socio-economic impacts of some selected agro-commodities – soybeans, palm oil, meat and coffee. A limited set of indicators, generally using existing (mostly) quantitative data, was used to assess these impacts. The study concluded that the loss of biodiversity in most cases did not result in the socio-economic improvements that are often assumed. The observed patterns were related to three different variables: differences between agro-commodities, between countries and between production regions. The authors judged the use of the method positively because the selected set of indicators offered insight into different sustainability impacts and their interactions, and allowed the use of both quantitative data and qualitative information. The number of indicators they used was limited but sufficient to build up a reliable and specific view.

*Source:* Kessler et al (2007)

nor do they help explain underlying causes of differentiated footprints between or within respective units of analysis.

## Conclusion

This chapter highlighted the many issues involved and the many challenges faced when defining and promoting sustainability in food provision. It also clearly identified two general trends. Over time, environmental and developmental concerns are becoming much more integrated, and sustainability is no longer a matter of agricultural production alone but an issue for the entire food-supply chain.

The process of globalization adds new challenges to promoting sustainability in food. As overall distances between food producers and consumers continue to grow, their connections through greater concentrations and increased flows of food intensify. This may accelerate the spread of certain risks, such as avian flu or swine fever, as well as threats to biodiversity. Another challenge is how to manage the impacts of new technologies that may spread easily without public authorities exerting much control. Globalization also makes clear that sustainability challenges relate to the production, processing, trade, as well as the consumption stages of food provisioning, and thus an integrated approach is needed.

Sustainability need not necessarily be translated into one particular system of supplying food and can be translated into various steps for the further promotion of sustainable provisioning at all levels (Sundkvist et al, 2005). For larger-scale food-production systems, this means that adequate institutions are needed to handle the geographical and temporal distances. To be effective, feedback mechanisms must be organized at the proper scale and operate at a time and rate that is appropriate to the stimulus that elicits the adjustment. Governments and markets have different strategies available to adapt production practices, to internalize the environmental costs of production in the price and to provide consumers with information about the full costs of producing a food product. Markets cannot set sustainability objectives by themselves; they require interplay and interrelationships with public-policy institutions (Buller and Morris, 2004) and civil-society organizations. Examples are environmental labels and environmental management schemes, showing the growing penetration and importance of private rules, conventions and market forms of regulation in sustainable food production that are facilitated by public policy through support schemes and payments and often initiated by pressure from consumers and NGOs.

Promoting sustainable food should be approached in an integrated manner, as sustainable food systems (Fresco, 2009). A sustainable food system is productive and responsive to changing demands, is resource efficient, puts explicit limits on emissions of GHGs and imposes strict energy efficiencies along the entire food chain, is responsive to changes in the opportunity costs of labour and allows for mechanization, ranges from producer to consumer and from farm to fork, and includes provision to reduce vulnerability.

---

*Take-home lessons*

- Sustainability has multiple dimensions and is becoming increasingly relevant in contemporary food provisioning.
- Different views on sustainability and how to realize it exist, which complicates determining a shared goal.
- No simple tools exist to measure sustainability in a holistic manner, but it is essential to take an integrated supply chain approach and not focus on only one stage of sustainability

---

## Notes

1 The total global area of arable and permanent cropland increased from 1.37 billion hectares in 1961 to 1.51 billion hectares in 1990 to 1.54 billion hectares in 2003. See: http://earthtrends.wri.org/searchable_db/index.php?theme=8&variable_ID=2333&action=select_countries (accessed 7 January 2011).
2 Some, for instance, have termed sustainability as combining the interests of 'people, planet and profit'.
3 Ostrom (1990) argues that the generalization suggested by Hardin should be replaced by a much more in-depth analysis of the concrete institutions developed by the users to manage common resources.
4 Past experiences also point to the limitations associated with using natural limits as the basis for policy interventions because predictions on environmental developments have repeatedly proven wrong.
5 The 'metabolic rift' (Foster, 1999) argues that the relatively closed-loop systems that existed in the past are now disrupted. In the traditional agrarian cycle, food was produced with the locally available resources, and wastes were reincorporated in the production system. Over the last three-quarters of a century, food producers and consumers have become increasingly separated by the growing rural–urban divide and the increased specialization of regions and countries, leading to more agricultural trade. Made possible by technological change, this process of distancing also fosters a highly productive, market-oriented and ultimately globalized commodity-producing system, rupturing the ecological relations of production. Agricultural inputs are progressively transformed (from organic resources to inorganic commodities), nutrient recycling reduced and new agronomic methods depend upon chemicals and bioengineered seeds and genetic materials produced under proprietary conditions. The increasing dependence on fossil fuels further constrains the future viability of industrial agriculture (McMichael, 2009).
6 The reduction of natural ecosystems is even more problematic when land is not used for producing food but for producing biofuels. The expanding area devoted to producing biofuels may lead not only to direct land-use change through cutting forests, but also to indirect land-use change when agricultural land is used for biofuels production and more land is needed for food production.

7 The water footprint is not included in this book. For further information, see Hoekstra andChapagain (2007).

## Further Reading

Carter, N. (2007) *The Politics of the Environment: Ideas, Activism, Policy*, University Press Cambridge, Cambridge: a general introduction on environmental policy.

Clay, J. (2004) *World Agriculture and the Environment: A Commodity-by-Commodity Guide to Impacts and Practices*, Island Press, Washington, DC: a detailed overview of the environmental impact from particular food crops.

Robbins, P., Hintz, J. and Moore, S. (2010) *Environment and Society: A Critical Introduction*, Wiley-Blackwell, Malden and Oxford: examines the relationships between environment and society.

## References

Aiking, H. and Boer, J. (2004) 'Food sustainability: Diverging interpretations', *British Food Journal*, vol 106, no 5, pp359–365

Atkins, P. and Bowler, I. (2001) *Food in Society: Economy, Culture, Geography*, Arnold Publishers, London

Barrett, C., Barbier, E. and Reardon, T. (2001) 'Agroindustrialization, globalization and international development: The environmental implications', *Environment and Development Economics*, vol 6, pp419–433

Boserup, E. (1985) 'Economic and demographic interrelationships in sub-Saharan Africa', *Population and Development Review*, vol 11, no 3, pp383–397

Buller, H. and Morris, C. (2004) 'Growing goods: The market, the state, and sustainable food production', *Environment and Planning A*, vol 36, pp1065–1084

Busch, L. (2000) 'The moral economy of grades and standards', *Journal of Rural Studies*, vol 16, no 3, pp273–283

Carson, R. (1962) *Silent Spring*, Houghton, Orlando

Carter, N. (2001) *The Politics of the Environment: Ideas, Activism, Policy*, Cambridge University Press, Cambridge

Clay, J. (2004) *World Agriculture and the Environment: A Commodity-by-Commodity Guide to Impacts and Practices*, Island Press,Washington, DC

Duchin, F. (2005) 'Sustainable consumption of food: A framework for analyzing scenarios about changes in diets', *Journal of Industrial Ecology*, vol 9, no 1/2, pp99–114

Ecosanres (2008) *Closing the Loop on Phosphorous*, SEI, Stockholm

Eddleston, M., Karalliedde, L., Buckley, N., Fernando, R., Hutchinson, G., Isbister, G., Konradsen, F., Murray, D., Piola, J., Senanayake, N., Sheriff, R., Singh, S., Siwach, S. and Smit, L. (2002) 'Pesticide poisoning in the developing world: A minimum pesticides list', *The Lancet*, vol 360, no 9340, pp1163–1167

Foster, C., Green, K., Bleda, M., Dewick, P., Evans, B., Flynn, A. and Mylan, J. (2006) *Environmental Impacts of Food Production and Consumption: A report to the Department for Environment, Food and Rural Affairs*, Manchester Business School and DEFRA, London

Foster, J. (1999) 'Marx's theory of metabolic rift: Classical foundations for environmental sociology', *The American Journal of Sociology*, vol 105, no 2, pp366–405

Fresco, L. (2009) 'Challenges for food system adaptation today and tomorrow', *Environmental Science & Policy*, vol 12, no 4, pp378–385

Gerbens-Leenes, P., Nonhebel, S. and Ivens, W. (2002) 'A method to determine land requirements relating to food consumption patterns', *Agriculture, Ecosystems and Environment*, vol 90, pp47–58

Greenberg, J. and Park, T. (1994) 'Political ecology', *Journal of Political Ecology*, vol 1, pp1–12

Hardin, G. (1968) 'The tragedy of the commons', *Science*, vol 162, no 3859, pp1243–1248

Helms, M. (2004) 'Food sustainability, food security and the environment', *British Food Journal*, vol 106, no 5, pp380–387

Hoekstra, A. and Chapagain, A. (2007) 'Water footprints of nations: Water use by people as a function of their consumption pattern', *Integrated Assessment of Water Resources and Global Change*, vol 21, pp35–48

Hoi, P. (2010) *Governing Pesticide Use in Vegetable Production in Vietnam*, WUR, Wageningen

Horrigan, L., Lawrence, R. and Walker, P. (2002) 'How sustainable agriculture can address the environmental and human health harms of industrial agriculture', *Environmental Health Perspectives*, vol 110, no 5, pp445–456

IFPRI (International Food Policy Research Institute) (2002) *Reaching Sustainable Food Security for All by 2020: Getting the Priorities and Responsibilities Right*, IFPRI,Washington, DC

Johnson, R. (2006) 'Sustainable agriculture: Competing visions and policy avenues', *International Journal of Sustainable Development & World Ecology*, vol 13, pp469–480

Kessler, J., Rood, T., Tekelenburg, T. and Bakkenes, M. (2007) 'Biodiversity and socio-economic impacts of selected agro-commodity production systems', *Journal of Environment & Development*, vol 16, pp131–160

Kirchmann, H. and Thorvaldsson, G. (2000)'Challenging targets for future agriculture', *European Journal of Agronomy*, vol 12, pp145–161

Konefal, J., Mascarenhas, M. and Hatanaka, M. (2005) 'Governance in the global agro-food system: Backlighting the role of transnational supermarket chains', *Agriculture and Human Values*, vol 22, no 3, pp291–302

Lang, T. (2010) 'Conclusion: Big choices about the food system', in G. Lawrence, K. Lyons and T. Wallington (eds) *Food Security, Nutrition and Sustainability*, Earthscan, London, pp271–287

Lawrence, G. (2005) 'Promoting sustainable development: The question of governance', in F. Buttel and P. McMichael (eds) *New Directions in the Sociology of Global Development*, Elsevier, Amsterdam, pp145–174

Leslie, D. and Reimer, S. (1999) 'Spatializing commodity chains', *Progress in Human Geography*, vol 23, no3, pp401–420

Leuck, D., Haley, M. and Harvey, D. (2004) *U.S. 2003 and 2004 Livestock and Poultry Trade Influenced by Animal Disease and Trade* Restrictions, USDA, Washington, DC

Malthus, T. (1798 [1970]) *An Essay on the Principle of Population*, ed. Antony Flew, Penguin, London

Maxwell, S. and Slater, R. (2003) 'Food policy old and new', *Development Policy Review*, vol 21, no 5/6, pp531–553

McMichael, P. (2009) 'A food regime genealogy', *Journal of Peasant Studies*, vol 36, no 1, pp139–169

Meadows, D., Meadows, D., Randers, J. and Behrens, W. (1972) *The Limits to Growth*, Universe Books, New York

Mol, A., Sonnenfeld, D. and Spaargaren, G. (eds) (2009) *The Ecological Modernisation Reader: Environmental Reform in Theory and Practice*, Routledge, London

Mrema, G., Baker, D. and Kahan D., (2008) *Agricultural Mechanization in Sub-Saharan Africa: Time for a New Look*, FAO, Rome, pp11–12

Nestle, M. (2002) *Food Politics. How the Food Industry Influences Nutrition and Health*, University of California Press, Berkeley

Oosterveer, P. (2007) *Global Governance of Food Production and Consumption: Issues and Challenges*, Edward Elgar, Cheltenham and Northampton

Ostrom, E. (1990) *Governing the Commons: The Evolution of Institutions for Collective Action*, Cambridge University Press, Cambridge

Picard, A. (2002) 'Pesticides banned many years ago still in some foods', *Globe & Mail*, Toronto, 15 October

Pillay, T. (1992) *Aquaculture and the Environment*, Fishing News Books, Oxford

Pimentel, D. (2009) 'Reducing energy inputs in the agricultural production system', *Monthly Review*, vol 61, no 3, pp92–101

Polanyi, K. (1944) *The Great Transformation*, Beacon Press, Boston

Pollan, M. (2007) 'Our decrepit food factories', *New York Times*, 16 December

Pretty, J., Brett, C., Gee, D., Hine, R., Mason, C., Morison, J., Raven, H., Rayment, M. and Van der Bijl, G. (2000) 'An assessment of the total external costs of UK agriculture', *Agricultural Systems*, vol 65, pp113–136

Pretty, J., Smith, G., Goulding, K., Groves, S., Henderson, I., Hine, R., King, V., Van Oostrum, J., Pendlington, D., Vis, J. and Walter, C. (2008a) 'Multi-year assessment of Unilever's progress towards agricultural sustainability I: Indicators, methodology and pilot farm results', *International Journal of Agricultural Sustainability*, vol 6, no 1, pp37–62

Pretty, J., Smith, G., Goulding, K., Groves, S., Henderson, I., Hine, R., King, V., Van Oostrum, J., Pendlington, D., Vis, J. and Walter, C. (2008b) 'Multi-year assessment of Unilever's progress towards agricultural sustainability II: Outcomes for peas (UK), spinach (Germany, Italy), tomatoes (Australia, Brazil, Greece, USA), tea (Kenya, Tanzania) and oil palm (Ghana)', *International Journal of Agricultural Sustainability*, vol 6, no 1, pp63–88

Rees, W. (2003) 'Economic development and environmental protection: An ecological economics perspective', *Environmental Monitoring and Assessment*, vol 86, no 1, pp29–45

Redclift, M. (2006) 'Sustainable development (1987–2005): An oxymoron comes of age', *Horizontes Antropológicos*, vol 12, no 25, pp65–84

Robbins, P., Hintz, J. and Moore, S. (2010) *Environment and Society: A Critical Introduction*, Wiley-Blackwell, Malden and Oxford

Roberts, J. and Grimes, P. (2003) 'World-system theory and the environment: Toward a new synthesis', in R. Dunlap (ed) *Sociological Theory and the Environment*, Rowman & Littlefield, Lanham, pp167–194

Roberts, J. and Hite, A. (eds) (2007) *The Globalization and Development Reader: Perspectives on Development and Global Change*, Blackwell, Malden and Oxford

Schumacher, E. (1973) *Small is Beautiful*, Blond Briggs, London

Sonesson, U., Mattsson, B., Nybrant, T. and Ohlsson, T. (2005) 'Industrial processing *versus* home cooking: An environmental comparison between three ways to prepare a meal', *Ambio*, vol 34, no 4/5, pp414–421

Strosser, P., Pau Vall, M. and Plötscher, E. (1999) *Water and Agriculture: Contribution to an Analysis of a Critical but Difficult Relationship*, EC, Brussels

Sundkvist, A., Milestad, R. and Jansson, A. (2005) 'On the importance of tightening feedback loops for sustainable development of food systems', *Food Policy*, vol 30, pp224–239

Tilman, D., Cassman, K., Matson, P., Naylor, R. and Polasky, S. (2002) 'Agricultural sustainability and intensive production practices', *Nature*, vol 418, no 6898, pp671–677

*Toronto Globe and Mail* (2002) 'Our decrepit food factories', *Toronto Globe and Mail*, 15 October

Unilever (2010) *Unilever Sustainable Agriculture Code*, Unilever, Rotterdam and London

Wackernagel, M. and Rees, W. (1996) *Our Ecological Footprint: Reducing Human Impact on the Earth*, New Society Publishers, Gabriola Island

WCED (World Commission on Environment and Development) (1987) *Our Common Future*, Oxford University Press, Oxford and New York

Williams, A., Audsley, E. and Sandards, D. (2006) *Determining the Environmental Burdens and Resource Use in the Production of Agricultural and Horticultural Commodities*, Cranfield University and DEFRA, Bedford

Xin, H. and Stone, R. (2008) 'Tainted milk scandal: Chinese probe unmasks high-tech adulteration with melamine', *Science*, vol 322, no 5906, pp1310–1311

# 4

# Regulating Food in the Global Network Society

This chapter aims to:

- illustrate the changing roles of governmental authorities in regulating food under global modernity;
- show how in reaction to governments' changing role, multilateral, private and civil-society-initiated food-governance arrangements are emerging;
- provide a conceptual background for these emerging alternative food-governance arrangements

## Introduction

Food's expanding international trade is profoundly transforming its regulation. In the past, regulating food was essentially based on national governments deciding to secure their economy and to feed their population with sufficient and safe food. But nowadays, these authorities are faced with highly increased volumes of imported food, while the actors and institutions involved in its production, processing, trade and consumption are integrated in changing transnational networks. Thus a key question is: how is food regulation organized – and how should it be organized – in the current context of global modernity?

While national governments' capacities to regulate food are declining, the demand for intervention seems to be only increasing. Unfamiliar agricultural and food risks, such as BSE and avian influenza H5N1, and, more recently, contamination of milk and other dairy products in China with melamine, draw public attention. These public concerns are further invigorated by the introduction of innovative technologies with unknown consequences for human and environmental health, such as with genetically modified (GM) food and nanotechnology. In reaction, consumers around the world are putting increasing pressure on authorities to address these challenges and to guarantee that their food has no adverse effects. Consumers today are concerned not only about the quality, safety and price of their food, but also about the health, social, ecological and animal welfare impacts occurring at all stages of the supply chain.[1]

Food needs to be regulated in such a way that these challenges can be effectively addressed. In recent years, we observe multiple innovative arrangements, including multilateral agreements, international institutions introducing specific prescriptions and private initiatives introducing food standards and labelling schemes. This chapter analyses these different arrangements and their respective backgrounds. In the discussion, particular attention is given to the legitimacy of private governance initiatives, as this is an often-criticized aspect of new global food-governance arrangements.

## The Nation State and the Regulation of Food

Current changes in food regulation occur against a background of a century of conventional food politics. Governments have engaged with food production and supply for centuries, but this engagement has intensified since the era of modernization, from the 19th century onwards. Growing urbanization and the start of the industrialization of food processing radically changed the relationship between producer and consumer and demanded stronger governmental control to balance the power difference between food producers and consumers. This phase of food regulation can be considered as being based on the conventional model of politics, as it entailed the regulation of food production, processing and trade through national governmental policy measures and institutions. Conventional food policy focuses on the nation state and the market, as they interact and (re)shape each other (Lang et al, 2009). States influence markets through public policy, laws, regulations and enforcement, while markets evolve and act to constrain state actions; at the same time, both are influenced by civil society. Authority is considered to be in the hands of a sovereign agency, and governing is essentially the relationship between the principal and subordinate state bodies responsible for implementing food-policy measures. Specialized agencies are expected to set food quality and safety standards on the basis of the latest results from scientific research and to oblige all private companies to adhere to these standards. To prevent their violation, special bodies are established for monitoring and control.

This approach to food regulation neatly fits into the dominant thinking in political science and practices in policy-making since the 1950s. According to this conceptualization, politics should be understood as a matter of rational and scientific management, building on functionalist models and resulting in a nation state-based model of policy-making. Such a policy process (including when dealing with food) had different, clearly distinguishable stages: formulation, policy-making, legislation, implementation, administration and adjudication (Lang et al, 2009). Despite its consistency, this model encountered several problems when used for analysing contemporary food politics. In recent decades, the role of the nation state is changing. While recognizing the large variations in how governments are organized, from centralized national states to different forms of federal government systems, similar trends can be observed in many parts of the world. Governments are no longer the sole centre of authority and control, entitled to make and enforce laws, and they have turned into more collaborative actors applying more indirect and softer forms of steering and involving various other societal actors, including private companies, in the process. To describe and analyse these changing ways of policy-making and the new roles governments take up, new concepts were introduced. Prominent among these is *governance*, which can be understood as 'the structures and processes that enable governmental and non-governmental actors to coordinate their interdependent needs and interests through the making and implementation of policies in the absence of a unifying political authority' (Krahmann, 2003, p331). By using governance and stakeholders instead of regulation and government, scholars explicitly refer to the shifting balance from top-down to more collaborative approaches to steering societal processes.

Present-day food politics is therefore no longer organized through government-based interventions only. National food provision includes larger ecosystems and wider social, political and economic networks. Food-related problems may then be the consequence of activities occurring at large distances, in both space and time, from the places of consumption. Increasingly, problems are cross-border and, even when they seem to have only a domestic character, they may ultimately prove insoluble within national borders (Wapner, 1998) (see Box 4.1 for an example).

National measures are often insufficient to solve food-related problems under the present food-production system. In the context of globalizing food provision, governments need to reconsider their position, particularly with regard to the division of tasks with non-governmental actors and between different levels of governing. Capturing the opportunities that trading high-value food products offer requires elaborate management systems and close control mechanisms 'from farm to fork' to meet the ever-more stringent food-safety requirements in consumer markets (Unnevehr and Jensen, 1999). Governments today can no longer determine their food quality and safety standards autonomously because global food trade requires their international harmonization

### Box 4.1 Avian influenza in the UK

In 2007, avian influenza was discovered in a large commercial turkey farm in Suffolk, UK. This was a unique case because no other incidents were reported in the surrounding areas, and further studies were conducted to determine where the disease originated. The most likely cause seemed the import of poultry products from farms in Hungary that were owned by the same company.

*Source:* De Krom (2010)

(Echols, 2001). Nation states and their constituent units are compelled to adjust to globally agreed-upon regulatory arrangements, 'because they will be excluded ... if they do not and, today, this means "economic marginalization and ruin"' (Schaeffer, 1995, p266). Despite this pressure, governments maintain considerable room for manoeuvring in the way they adapt to these global requirements, although they cannot ignore them completely.

Ecological, political and economic interdependence thus scrambles the logic of national sovereignty (Wapner, 1998). National authorities are increasingly sharing responsibilities with various supply-chain actors, such as food-processing firms and retailers (Marsden et al, 2010). Intervening effectively becomes difficult for governments, as this requires the involvement of multiple (inter) national, regional and local authorities (both public and private), each with their particular assignments and regulatory instruments. To be effective in their interventions, these different actors must synchronize their actions. Private firms and civil-society actors are increasingly involved in responding to consumer demands for food safety and sustainability. These responses, in turn, have been significantly influenced by the transnational character of food supply chains, with food-related firms empowered through market concentration. These firms can use their increased power resulting from market concentration (Boström and Klintman, 2008; Hatanaka et al, 2005). Nowadays, nation states struggle to deal not only with the tension between state and market, but also with the de-territorialized and decentred mobilities of the numerous flows of food in the global network society.

Despite recent institutional reforms in the regulation of food (Lang and Heasman, 2004), individual nation states no longer have the capacity to adequately respond to contemporary food problems without relying on non-state actors (Held et al, 1999; Held, 2004). For instance, the Dutch food safety agency is legally obliged to test spinach imported into the Netherlands for the presence of nitrate. In 2004 the agency was able to test only nine samples of all imported fresh spinach (VWA, 2005). To assure the safety of food, retailers and producers therefore must fill in the gaps left by national authorities. As a consequence, the authority of individual nation states is threatened and their interventions risk becoming delegitimized because norms and standards seem to evolve at global rather than national levels. Moreover, there are no particular mechanisms for setting such norms and standards, as they may emerge from NGOs, private companies or governments or from various combinations of these (Oosterveer, 2007). Global change, transnational flows and multilateral agencies become 'original' sources of normative claims, rather than just serving as complements or counterparts of nation states (Spaargaren and Mol, 2008). The increasing globalization of food then results in a shift away from governance by national governmental authorities towards responsibilities taken by other actors; this trend notwithstanding, national food regulatory agencies are as critically important today as in the past.

Various scholars challenge this conclusion. Cheshire and Lawrence (2005, p45), for instance, argue that: 'the nation state has not so much "lost" power to local or global actors as has found new, advanced liberal ways of governing through a functioning network of state and non-state actors'. In their view, there is no hollowing out of the nation state through globalization, because it remains the cornerstone of international affairs. Besides, globalization is not stripping every nation state from its long-standing sovereignty in equal measure because of differences in power, as 'state and non-state actors strive to establish principles of their choosing, using the mechanisms they have at their disposal' (Drahos and Braithwaite, 2001, p124). These commentators argue that despite the significant changes that are taking place worldwide, the nation state is still the most complete nexus of relationships that exists within the international order. Nation states provide the required stable legal context in which private companies can operate and NGOs can become involved (Tansey and Worsley, 1995; Sassen, 2006). It is only within this legal and political framework that it is possible (at least to a certain extent) to balance the economy with other interests, although one should be conscious that 'international trade is a transnational reality and those seeking to promote sustainable development must seek to operate within it' (French, 2002, p139). The nation state remains recognized by many to be a vital institution possessing important symbolic and real authority (Polanyi, 1944). In matters of food, most citizens still hold their national governments responsible for addressing their concerns and for responding appropriately to newly emerging crises. These commentators underline the continued importance that should be attached to the position of nation states in food governance, but they cannot ignore that nation states operate in a fundamentally different context. So, while maintaining their relevance for food governance, the national authorities now must deal with regulating food in cooperation with other states and various social actors.

Hence, contemporary food governance in Western countries has become an increasingly 'interactive process of state and public laws and policy with private interests and actors. The private sector governance may be corporate led or originate from civil society' (Lang et al, 2009, p81). This interactive process can be consensual but also conflictual, and different contingent global food-governance regimes may emerge. We can distinguish the following models:

- international governmental collaboration;
- international governance regimes;
- global governance arrangements.

International governmental collaboration takes place when independent states work together to find a solution for a particular problem. More intensive collaboration between independent states may result in the creation of international governance regimes through multilateral institutions, such as the UN or WTO.[2] Finally, when strict distinctions between domestic and foreign policy issues, and between governmental and non-governmental actors disappear, various global governance arrangements may arise. In this way, more regulation is organized through markets and civil-society involvement.

International collaboration takes place when independent states work together to find a solution for a particular problem (e.g. an animal health problem such as foot-and-mouth disease) that may threaten their food provision (Burgiel et al, 2006). Governments may choose to collaborate in such instances, but they maintain their independence and may terminate the collaboration whenever they consider it appropriate to do so. In such instances, governments behave as rational actors.

This form of incidental collaboration among different governments has been practised for centuries and therefore cannot be considered very innovative. The introduction of international food-governance regimes and global governance arrangements deserves more attention. Therefore, below we develop them further and introduce the WTO as an example of an international regime relevant for global food provision, after which we discuss various global food-governance arrangements.

## International Food Regimes: The Case of the WTO

A thickening web of multilateral agreements and institutions (in)directly involved in food governance is being established, constituting global food-governance regimes (Young, 2000; Drahos and Braithwaite, 2001). These global regimes effectively strengthen the capacity of individual nation states (even while usurping their power) to control food quality, safety and trade. Among them we find the science- and government-based institutions Codex Alimentarius, the International Office of Epizootics (OIE), the International Plant Protection Commission (IPPC) (Lee, 2009) and the European Food Safety Authority (EFSA). With its high profile, the WTO stands out, however (Paterson, 1999; Griffin, 2003).

The WTO was established in 1995, as the successor to the General Agreement on Tariffs and Trade (GATT), which originated in 1945 when it was founded to harmonize international trade (Narlikar, 2005). The WTO, is a membership organization with more than 150 member states and that includes the EU as a multilateral member, while scores more countries negotiate their future accession. The WTO's objective is to reform and liberalize international trade policies in order to ensure economic growth, development and employment. Achieving this goal means reducing and eliminating unnecessary and unjustified trade barriers, while respecting the sovereignty of other member states. The WTO negotiates compact deals that cover various elements through multi-annual negotiation rounds. Within a few years, the organization reached a dominant position in the regulation of global trade, at least partly because, as one of very few multilateral institutions, the WTO has installed an internal dispute-settlement mechanism through which one member can file a complaint against another member on a specific trade-related issue. Compliance to the final judgement is enforced, as the WTO permits the member whose complaint is judged as justified to suspend certain trade concessions to the violating country, typically by raising tariff rates on that country's exports to such a level that the damage from the loss in exports is compensated (Kastner and Pawsey, 2002).

On food, the WTO's arrangements are built on national governmental regulations and supplemented by specific guidelines for international trade. These arrangements generally tend to avoid placing limits on nation state sovereignty, and thus governments are not allowed to interfere with internal policies of other member states. Through multilateral negotiations, the WTO intends to create an 'even playing field' for competition among companies from different countries. The WTO allows its members to intervene in international trade only on the basis of product-related 'process and production methods' (ppms), as well as on the product's properties, but not on the basis of non-product-related ppms (Charnovitz, 2002). This means that production methods may not be subjected to import requirements unless they have a verifiable impact on the end-product. Food safety can be incorporated into trade regulations only when measures are intended to secure public health and safety, are based on sound scientific evidence and are limited to objectively verifiable product characteristics. So, according to this approach, whether or not a food product is produced or captured sustainably should not be a condition in international trade regulations. In this way, the WTO intends to prevent the imposition of barriers to trade that would advance the countries with more developed governmental institutions. Evidently, this approach to trade regulation is aimed at securing and facilitating free and global flows of food and not at guaranteeing the sustainability of the food produced (Potter and Burney, 2002).

For particular problems in trading food, two separate covenants within the WTO framework are relevant, the Sanitary and Phytosanitary (SPS) agreement and the Technical Barriers to Trade (TBT) agreement. The SPS agreement was concluded as part of the Agreement on Agriculture within the WTO in 1995. The SPS agreement allows governments to intervene in trade to assure food safety, to protect animal and plant life and to secure human health from risks arising from the spread of pests and diseases and from additives, contaminants and toxins in food. All measures must be based on scientific evidence about the risks involved, while precautionary measures are allowed only temporarily, until scientific proof is available. As much as possible, measures should be harmonized via the international standard-setting bodies: the Codex for food safety, the OIE for animal health and the IPPC on plant health. Thus 'the agreement recognizes the governmental and national interests in food safety regulation ("appropriate level of protection"), while significantly abridging sovereignty and rejecting cultural differences as a basis for regulation' (Echols, 2001, p93).

The TBT agreement had already been adopted in 1979 to avoid unnecessary barriers to trade in agricultural and industrial products through the use of (mandatory) technical regulations and (voluntary) standards. To achieve this, members encouraged internationally agreed-upon standards and regulations, such as those developed by the International Organization for Standardization

(ISO); they also applied the principles of *equivalence* (accepting other countries' technical regulations) and *mutual recognition* (recognizing the results of other countries' assessment procedures). The TBT agreement allows member states to apply their own national standards and regulations, even when they differ from international ones, as long as they are based on sufficient scientific evidence, only concern product-related characteristics and fulfil the criteria of objectivity, proportionality and non-discrimination among the member states. Applying the agreement should result in WTO members treating imported products no less favourably than 'like products' that are produced domestically. In practice, this objective remains difficult to translate into actual measures (see Box 4.2 for the example of the dolphin–tuna dispute).

The TBT agreement has evolved into the most relevant regulatory guideline for governmental authorities when developing and introducing eco- and other labels for food. WTO members must ensure that their regulations and standards, including eco-labels, are not more trade restrictive than necessary to fulfil a legitimate objective (Appleton, 1999; Motaal, 1999). Yet, developed countries impose increasingly higher and more complex quality and health and safety requirements on exporting countries, which developing countries regularly find difficult to meet.[3] It seems that only a thin boundary exists between 'legitimate' measures taken to safeguard consumers and the environment and those that merely protect national interests. Developing countries fear that such certification requirements are yet another barrier for entry into the lucrative food markets of Western countries. These countries repeatedly threaten to challenge such requirements before the WTO, but so far they have intervened in only isolated cases.

Thanks to the SPS and TBT agreements, the WTO constitutes a global food regime that promotes international trade in food. At least in part, however,

## Box 4.2 The dolphin–tuna case

The dolphin–tuna dispute started in 1988 when an environmental group in the US sued its federal government because it did not implement its legal obligation to curtail the incidental killing of marine mammals in commercial fisheries. The NGO claimed that Mexican fishermen were killing dolphins in the eastern tropical Pacific Ocean, where dolphins swim with tuna. The federal court agreed that the administration was not upholding the law and ordered Mexican tuna imports to be banned from the US. Mexico argued that its right to sell tuna in the US had been violated and asked the GATT to adjudicate the matter. The panel concluded that the US was in violation of its GATT obligations, but Mexico decided not to pursue the case further and the panel report was never formally adopted.

The case was put to rest until 2008 when the Mexican government argued that the US requirements for labelling canned tuna as 'dolphin safe' were unfairly discriminatory against its fishermen using large nets to encircle tuna ('purse-seine' fishermen), because their new techniques allowed them to capture tuna without killing any dolphins. On this basis, Mexico requested the establishment of a WTO panel in March 2009 to rule on their complaint. The case is still not closed because US authorities proposed to put the case to another multilateral organization, the North American Free Trade Association (NAFTA).

*Source: Bridges Weekly* (several issues)

because of the problems that occur when trying to include social and environmental impacts in trade regulations, social actors are dissatisfied and look for other means to achieve this aim.

## Multi-actor Collaboration

Several NGOs are opposing the establishment of 'global regimes of food production and trade' through the WTO, because they consider food to not be an issue for economic concern alone. Therefore, in addition to multilateral government-based institutions, NGOs and private companies also engage in regulating food trade. NGOs may exercise public pressure and take initiatives to cover broader issues in the regulation of trade in food, sometimes with high levels of effectiveness (Dicken et al, 2001). The rising efficacy from a multitude of these social movements contributes to the permanent reformulation of objectives and practices within global food governance (Speth, 2002). Official governmental regulations are increasingly complemented by governance arrangements introduced by private firms, such as the certification schemes ISO 14001 and HACCP (Unnevehr and Jensen, 1999), and by NGOs, for example, those certifying fair trade (Zaccaï, 2007) and organics (DuPuis and Gillon, 2009). These new multi-level, non-hierarchical, information-rich, loose networks of institutions and actors presently drive global food governance (Stripple, 2006) because they prove more effective, flexible and responsive and promote social learning (Lemos and Agrawal, 2006). Societal self-regulation involves particular policy instruments, actor constellations and institutional settings, or 'innovative global governance arrangements' (Treib et al, 2007). Such arrangements are blurring previous distinctions, such as those between the public and private spheres and between the national and international levels in policy-making. Governing food in global modernity is no longer the sole responsibility of governmental authorities but involves transnational and national political institutions, together with market actors and NGOs. 'World politics is being transformed into a "polycentric" or "multinucleated" global political system operating within the same geographical sphere (and/or overlapping spaces)' (Cerny, 1999, p190). Various multi-level, multi-actor governance arrangements (Van Tatenhove et al, 2000) are emerging, combining changes at the supply side (e.g. increased control within food-supply chains) with adaptations at the national and international levels (such as legal initiatives and private certification). Governing food is therefore necessarily more complicated in the globalizing world than under conditions of the national governments-based system of the past because it involves a wider variety of social actors and deals with more issues and a greater variety of them (Buttel et al, 2006). These global food-governance arrangements have not evolved into a stable system because new actors and new items repeatedly become included.[4] They can, nevertheless, be regarded as temporary stabilizations of the content and organization of global food provision as 'a policy domain' (Arts et al, 2006).

## Box 4.3 Marine Stewardship Council

Overfishing is risky because it reduces existing fisheries stocks and their natural regeneration capacity, inevitably leading to lower catches in the future and threatening the present and future livelihoods of millions of subsistence and commercial fishermen. In 1997, the diversified, transnational firm Unilever and the environmental NGO World Wide Fund for Nature (WWF) founded the MSC, assuming that all actors involved in a fishery share an interest in securing its future. MSC later became an independent, global not-for-profit organization responsible for certifying sustainable fisheries through its own (MSC) label. The MSC Principles and Criteria for Sustainable Fishing are used as a standard in a third-party, independent and voluntary certification programme. The standard is based on three principles. First, the fishery must not lead to over-fishing or depletion of the exploited fish population. Second, the fishery should allow for the maintenance of the ecosystems on which it depends. Third, the fishery should have in place an effective management system that respects local, national and international laws and standards. These principles are translated by a certifying agent into a sustainable fisheries management plan together with all actors concerned, as they agree on the amount, the way and the timing of catching fish, as well as on the implementation of particular accompanying measures. The certifier must be accredited by MSC first, before they are allowed to perform this task.

*Source:* Oosterveer (2008); see also Chapter 7

Private companies and NGOs are involved in market-based and voluntary forms of governance, such as voluntary agreements, standards and certification, eco-labelling and informational systems (Boström and Klintman, 2008). These emerging cross-scale governance arrangements create opportunities for hybrid management options: *co-management* (between civil society and state), *public–private partnerships* (between state and market) and *private–social partnerships* (between civil society and the market) (see Box 4.3 for the Marine Stewardship Council as an example of a private–social partnership).

Private companies constitute another group of societal actors engaged in initiating global governance arrangements when they try to secure the quality and sustainability of their product supply, or to strengthen their position within the supply chain or on the food market. A leading example of this is the introduction of GlobalGAP as a set of worldwide standards for sustainable and safe food (see Box 4.4 for details).

The success in terms of both the quantity and impact of these private global food-governance arrangements is astonishing. Nevertheless, as innovative governance arrangements, they need to obtain public legitimacy and authority to be effective. Here a distinction should be made between input legitimacy, through an official mandate, and output legitimacy, through concrete results being accepted (Skelcher, 2009). Before discussing the legitimacy challenge of private governance arrangements more specifically, it may be helpful to take a step back and examine three conceptual frameworks developed for analysing global multi-actor governance arrangements.

## Box 4.4 GlobalGAP

GlobalGAP, established as EurepGAP[5] in 1997 by the Euro-Retailer Produce Working Group (EUREP), combines food safety standards with sustainable food production measures (Van der Grijp and Hond, 1999; Campbell, 2005). Prior to the establishment of GlobalGAP, European retailing firms reacted to food-safety incidents individually, leading to confusion and competition. Larger European retailers then recognized the need for further harmonization and chose to promote 'integrated crop management'. Their objective was to develop common, verifiable environmental and food-safety standards that would be accessible for mainstream farmers (Gawron and Theuvsen, 2009). For this purpose, they developed 'good agricultural practices' to assure the safety of food in combination with on-farm integrated production systems to guarantee environmental sustainability. At first, GlobalGAP focused on fresh fruits and vegetables only, but it later expanded to include other products (Humphrey, 2008). GlobalGAP identifies four key topics: food safety; environmental protection; occupational health, safety and welfare; and animal welfare. It is a private, voluntary process standard based on third-party certification, and as a business-to-business standard, it does not label food products to inform consumers. GlobalGAP protocols have become very popular and are applied by nearly all leading European retail chains, as well as by many farmers and auditing and agro-input firms from different continents. According to its own website, the organization has issued over 93,000 certificates to producers in more than 100 countries. The initiative expands by including more members and more food products and by benchmarking its standards against other international food safety standards. GlobalGAP also actively promotes benchmarked national standards that are regarded as equivalent. This makes it possible to adapt GlobalGAP's own protocols to national circumstances while continuing to recognize the national standard's value.

*Source:* Campbell (2005); Humphrey (2008); www.globalgap.org (accessed 15 April 2010)

## Conceptualizing Multi-actor Governance

Contemporary global food governance includes multiple layers, from local to global, and multiple actors, from private firms to non-governmental interest groups (Mol, 2001). The undermining of the state under conditions of globalization does not leave the state irrelevant, as it is also through the state and its institutions that globalization materializes (Spaargaren and Mol, 2008). The de-territorialization of authority and its shift from state to market still imply that such authority must be shaped, channelled and enabled by institutions and networks that remain overwhelmingly rooted in (the territory of) nation states (Mol, 2010). Nation states are thus not disappearing but transforming (Castells, 1996) and becoming part of a complex policy network with power-sharing and negotiated decision-making processes. The involvement of other, non-state actors results in the 'fluidization' of regulatory practices (Lipschutz and Fogel, 2002; Sassen, 2006). In particular, global civil society is 'constantly active – with ambivalent results – in redirecting global economic processes into less harmful directions' (Mol, 2001, p116) and has become an important political actor 'in global governance by mobilizing means of governance that operate independently of the state system' (Wapner, 1997, p81). Under these circumstances, 'soft' governance instruments, such as labels, standards and certification schemes seem to be increasingly preferred over 'hard' instruments, such as bans, moratoriums, quotas or other legal requirements (Klintman and

Boström, 2004). These innovative governance arrangements should be analysed with conceptual tools other than those of conventional regulation. Next, we present three potential explanatory frameworks that offer competing views on this transition.

First, Sassen (2006) argues that having completed the phase of the nation state, we are now entering an era with a different assemblage of territory, authority and rights. This new organizing logic, based on global configurations, is already dominant, but the nation state remains relevant because its formalization and institutionalization are the most developed. The nation state also remains one of the key enablers and enactors of the emergent global scale of governance. In the organizing logic of the global, the diverse emergent assemblages of territory, authority and rights have one systemic feature which is that they are denationalized. A typical contemporary structuration of territory, authority and rights, external to the nation state, includes the private institutional order linked to the global economy. Sassen acknowledges that global governance arrangements may emerge from capabilities present among key actors, such as private companies, so there is no need to include all stakeholders, which is a necessary feature of state-based regulation. Braithwaite (2008) observes that non-state regulation has grown more rapidly than state regulation, so he claims that it is best to conceive our era as one of *regulatory capitalism*. The corporatization of the world is both a product of regulation and the key driver of regulatory growth. Therefore, the coherent nation-based food regulation of the past is replaced by multiple arrangements, whereby nation states provide only the general regulatory background, while central principles, procedures and responsibilities are in the hands of non-state actors. Key questions then become how these emerging arrangements are made possible, and on what their authority and legitimacy is based.

To answer these questions, Sassen argues that authority today is not necessarily based on the state's formal, legal position; rather, it has a more social character. Or, as Litfin (2000, p123) argues, 'authority should not be confused with power or control ... Authority in the Western tradition of political thought is construed as the rightful governance of human action by means other than coercion or persuasion.' Therefore, if authority is less formal and instead social, its legitimacy is based more on acceptance than on coercion.

A second view was developed by Rosenau (2007), who sees the loci of authority as having been shifted by globalization. States are weakening and although territory remains important in peoples' lives, the state's role is more flexible and less binding than in the past. This results in a *bifurcation* away from the nation state, whereby authority is relocated 'upward' towards transnational organizations and 'downward' towards sub-national groups. Van Kersbergen and Van Waarden (2004) add horizontal transfers to this explanation, within the public sector towards the judiciary institutions and from public to semi-public organizations. In the same vein, they also identify mixed, vertical–horizontal shifts, such as those from national public to international private regulation.

States are not oblivious to these changes, although it is not their sovereignty or jurisdiction that becomes problematic; rather, the problems stem from exclusivity and the scope of their competence within these multi-centric systems of governance. According to Rosenau (2007), national governments are becoming one level and one actor amidst a plurality of arrangements with multiple levels and multiple actors. This is referred to as *multi-level, multi-actor governance*. In this context, Rosenau argues that authority is becoming de-territorialized and that multiple 'spheres of authority' are emerging, which refer to the different entities wherein authority is presently located. A sphere of authority is determined by the issuance of directives from its leadership and compliance by its adherents. Politics in these spheres of authority are more chaotic and informal than in the nation state because there are not only conventional actors and governance levels but also unconventional ones where each actor must redefine and construct his own role. Under these conditions, authority has to rely on acceptance from the people involved. If people revoke their compliance to particular requirements, authority is directly undermined.

A third perspective is developed by Castells, who argues for a global network society. Castells (1996) reintegrates the functional unity of different elements at distant locations. This ongoing process of time–space distantiation links social practices at different (sometimes distant) locations. Today's globalized food provision clearly fits this perspective. Still, although global agri-food networks and flows increasingly dominate contemporary food provision, this does not mean there is a one-dimensional process of transfer of power from local and national levels to global levels; the transfers of power sometimes conflict, and so the outcomes are not already known (Mol, 2001). The complexity of global food flows and networks prevents unilateral action to exercise determining control over their direction and structuration. At the same time, global food flows should not be reified because they are always open to change via interventions from purposeful social actors (Gille, 2006).

The simultaneous presence of globalized food-provision systems and localized food production and consumption practices may lead to unexpected consequences requiring governance interventions that can handle the 'contrasting logic between timelessness, structured by the space of flows, and multiple, subordinated temporalities, associated with the space of places' (Castells, 1996, p468). This means that more is needed than simply adding a new category of global-level governance instruments to the presently available local and national ones. The ongoing interdependence, exchange and interrelationships between the different spheres of structuring time and space, demands an integrated governance approach that does not separate the global and the local but that combines them. Therefore, global food governance has to address the material, financial and informational flows and social networks at the global level and, at the same time, relate these global food flows and networks to the specific local environmental and social impacts of producing, processing, retailing and consuming food. Overall, Fulponi (2006) observes three trends: (1) a move towards private, voluntary quality and safety systems that include process as well as product characteristics, (2) the emergence of global coalitions for setting standards and (3) the increased use of global business-to-business standards.

Consequently, the multiplying demands for assurance about the safety and quality of food, and about its social and environmental performance, spurred the retail sector and international NGOs to develop standards and assurance schemes and to translate them back into the food-supply chain.

Sassen, Rosenau and Castells all agree that globalization has fundamentally altered the position of the nation state in present (food)-governance arrangements. They have lost their privileged position, but it is not clear how to conceptualize their present position. While Castells is essentially arguing for the growing irrelevance of national authorities, Rosenau still considers them as pivotal in governing global dynamics, while Sassen occupies a middle position by incorporating nation states in multiple assemblages of territory, authority and rights. Nevertheless, they all agree that non-state actors are much more actively engaged with contemporary food-governance arrangements, but this transition raises new questions on legitimacy and effectiveness.

## Changing Roles of Societal Actors in Food Governance

The growth in the number of innovative arrangements is closely related to the changing roles of societal actors involved in global governance. In particular, NGOs, consumers and private companies are taking up unprecedented responsibilities.

Civil-society organizations, such as environmental NGOs and consumer groups, are actively searching for improved global food-governance arrangements, resulting in the introduction of various non-exclusive, non-hierarchical, post-territorial, adaptive and flexible instruments (Goverde and Nelissen, 2002; Karkkainen, 2004). Networks of NGOs and globally shared civil-society norms and values facilitate private 'counter-authority' against the power, logic and authority of the global market (Mol, 2010). As members of the WTO, governments are seriously restricted in how they can include environmental considerations in their trade regulations, but NGOs and private companies are not, and hence they have much more room for intervening.[6]

In doing so, they complement, and sometimes, as shown in the case of the MSC (see Box 3.3), even replace nation states in regulating global food provision. More than governments or private companies, NGOs are capable of building trust among the different actors involved despite the sometimes large distances in time and space (Oosterveer and Spaargaren, 2011). Global food-governance arrangements, such as certification and labelling schemes, establish previously non-existing connections between societal concerns and social actors (Schaeffer, 1995; Dicken et al, 2001). As (particularly in Europe) consumers trust NGOs rather than governments or private companies, NGO-initiated labelling schemes can respond to the growing need for reliable information among consumers (Raynolds, 2000). Such schemes offer transparency about the production process, allow traceability of the food product throughout the supply chain and verify for the consumer the standards and criteria applied. Information offered in this way is not necessarily limited to product-related characteristics; it can also address other producer and consumer concerns. The rapidly increasing demand for products labelled according to several such NGO-initiated schemes – organic food, free-range eggs, fair trade coffee and MSC-labelled fish (Oosterveer, 2006) – underlines their growing influence in global food governance.

These 'non-state and market-driven' instruments are not based on formal governmental decision-making procedures, and thus the allocation of authority and power within them is not always transparent, and neither is their legitimacy. Nevertheless, authority (as a fusion of power with legitimate social purpose) clearly exists in private markets as well as the public domain of the state. Non-state authority has been gaining ground since the 1990s, and its democratic accountability seems to have been constructed through such mechanisms as representation, deliberative interactions, transparency and accountability (Mol, 2010). Private initiatives often 'appear to have been accorded some form of legitimate authority' (Hall and Biersteker, 2002, p4). This legitimacy is built not so much through procedures, such as democratic elections for nation states, but through their (intended) output (Bernstein et al, 2009). NGOs argue that they are contributing to a recognized public goal, such as environmental protection or securing food safety, which national governments cannot sufficiently guarantee. They can also base their authority on broad participation from stakeholders and transparency in the decision-making process. Furthermore, the use of scientific methods and data in the implementation process is regarded as an important justification for the initiative's authority. Surprisingly, this diffuse basis of authority and legitimacy does not seem to be problematic in practice (Gulbrandsen, 2004). On the contrary, the absence of formal authority is used to justify NGOs' legitimacy precisely because this precludes the imposition of procedures without open consultation and widespread public consent.

## Conclusions

For more than a century, conventional food regulation was based on sovereign nation states making decisions to protect their populations and the environment. Today, accelerated globalization challenges this practice, and the role of the nation state and governmental authorities has fundamentally changed. Today, most governments cannot effectively control international food trade because of its size and because production areas, structures of production and consumption constitute swiftly changing transnational networks. Moreover, it is increasingly hard to distinguish among issues that are national, international or global.

The ongoing process of globalization, described in Chapter 2, has created room for other societal actors to engage in the governing of food provision. Governments are bound, in part, by multilateral agreements into which they have entered to promote international trade and to secure a more-level playing field for international competition. Governmental tasks have been taken over to some extent by private societal actors, including NGOs and private companies.

As a result, food governance today involves (sometimes shared and sometimes competing) responsibilities among national governments, multilateral regimes, local states and a multitude of private initiatives. Many such multi-actor food-governance arrangements are introduced at the global level, or at least have cross-border impacts. Consequently, there are often competing regulatory frameworks, whereby NGO-initiated arrangements are continuously growing in number. In these new arrangements, national governments continue to play a critical role, but other actors, including food retailers, consumers and NGOs, play increasingly important roles as well. For many, this is a confusing situation that requires further control and harmonization. In practice, this is not a simple process because there is no recognized authority that can exercise this role. At the same time, however, the presence of various governance arrangements increases opportunities to address particular concerns that are not (yet) part of national regulation. In particular, in matters of sustainability and cross-border impacts from food provision, multilateral and private governance arrangements seem to offer promising ways to move forward.

---

## *Take-home lessons*

- Governmental authorities are no longer able to effectively regulate international food trade.
- New global food-governance arrangements have emerged through collaboration between governments, the creation of multilateral agreements and institutions, and different private initiatives from corporations and NGOs.
- These global food governance arrangements develop and apply innovative instruments such as certification and labelling of food producers and food products.
- Privately initiated food-governance arrangements are faced with the challenge of securing their legitimacy because, in contrast to national governments, they cannot rely on well-accepted procedures.

---

## Notes

1 Not all consumers are concerned about all impacts all of the time, but most consumers are concerned about at least some impacts at some moments in their life (Kjaernes et al, 2007; Oosterveer et al 2007).
2 Note that the concept of regime is used here with a somewhat different connotation than in the case of food regimes introduced in Chapter 2.
3 See Otsuki et al (2001) for an interesting case on the impacts of introducing a food safety standard for the presence of aflatoxins on nuts imported into the EU.
4 One recent issue that is included is the global warming impact from food production and trade. See Chapter 5.
5 The renaming occurred in 2007 to underline the global aspirations of the initiators (Henson et al, 2009).
6 Some WTO members, particularly developing countries, argue that governments should also intervene in private standardization and labelling schemes because these may be trade restrictive as well. They claim that such schemes create higher producer costs and impose requirements that are often in excess of agreed-upon international standards and not adapted to local conditions. This issue has been debated within the SPS Committee of the WTO, but so far disagreements among the different member states prevail, first on the legal authority to become engaged in this matter at all and second on how to settle disagreements on private standards in the case that they did intervene (*Bridges Weekly*, 2010).

## Further Reading

Narlikar, A. (2005) *The World Trade Organization: A Very Short Introduction*, Oxford University Press, Oxford: presents the WTO and its main agreements.

Braithwaite, J. (2008) *Regulatory Capitalism: How it Works, Ideas for Making it Work Better*, Edward Elgar, Cheltenham and Northampton: discusses governing in the context of globalization.

Marsden, T., Lee, R., Flynn, A. and Thankappan, S. (2010) *The New Regulation and Governance of Food: Beyond the Food Crisis?*, Routledge, New York and London: discusses contemporary food governance.

Lang, T., Barling, D. and Caraher, M. (2009) *Food Policy: Integrating Health, Environment and Society*, Oxford University Press, Oxford: examines changing practices in food policy.

## References

Appleton, A. (1999) 'Environmental labelling schemes: WTO law and developing countries implications', in G. Sampson and W. Chambers (eds) *Trade, Environment, and the Millennium*, United Nations University Press, Tokyo, pp195–222

Arts, B., Leroy, P. and Tatenhove, J. (2006) 'Political modernisation and policy arrangements: A framework for understanding environmental policy change', *Public Organization Review*, vol 6, pp93–106

Bernstein, S., Clapp, J. and Hoffmann, M. (2009) *Reframing Global Environmental Governance: Results of a CIGI/CIS Collaboration*, CIGI, Waterloo

Boström, M. and Klintman, M. (2008) *Eco-standards, Product Labelling and Green Consumerism*, Palgrave MacMillan, Houndmills

Braithwaite, J. (2008) *Regulatory Capitalism: How it Works, Ideas for Making it Work Better*, Edward Elgar, Cheltenham and Northampton, MA

*Bridges Weekly* (2010) 'Private standards, mediation cause conflict at WTO committee meeting', *Bridges Weekly*, vol 14, no 25, 7 July, ICTSD, Geneva

Burgiel, S., Foote, G., Orellana, M. and Perrault, A. (2006) *Invasive Alien Species and Trade: Integrating Prevention Measures and International Trade Rules*, CIEL and Defenders for Wildlife, Washington, DC

Buttel, F., Spaargaren, G. and Mol, A. (2006) 'Epilogue: Environmental flows and early twenty-first century environmental social sciences', in F. Buttel, G. Spaargaren and A. Mol (eds) *Governing Environmental Flows in Global Modernity*, The MIT Press, Massachusetts, pp351–369

Campbell, H. (2005) 'The rise and rise of EurepGAP: European (re)invention of colonial food relations?', *International Journal of Food and Agriculture*, vol 13, no 2, pp1–19

Castells, M. (1996) *The Rise of the Network Society. Volume I of The Information Age: Economy, Society and Culture*, Blackwell, Malden and Oxford

Cerny, P. G. (1999) 'Globalization, governance and complexity', in A. Prakesh and J. Hart (eds) *Globalization and Governance*, Routledge, London and New York, pp188–212

Charnovitz, S. (2002) 'Solving the production and processing methods (PPMs) puzzle', in K. Gallagher and J. Werksman (eds) *The Earthscan Reader on International Trade and Sustainable Development*, Earthscan, London, pp227–262

Cheshire, L. and Lawrence, G. (2005) 'Re-shaping the state: Global/local networks of association and the governing of agricultural production', in V. Higgins and G. Lawrence (eds) *Agricultural Governance: Globalization and the New Politics of Regulation*, Routledge, London and New York, pp35–49

De Krom, M. (2010) *Food Risks and Consumer Trust: European Governance of Avian Influenza*, WUR, Wageningen

Dicken, P., Kelly, P., Olds, K. and Yeung, H. (2001) 'Chains and networks, territories and scales: Towards a relational framework for analysing the global economy', *Global Networks*, vol 1, no 2, pp89–112

Drahos, P. and Braithwaite, J. (2001) 'The globalisation of regulation', *The Journal of Political Philosophy*, vol 9, no 1, pp103–128

DuPuis, E. and Gillon, S. (2009) 'Alternative modes of governance: Organic as civic engagement', *Agriculture and Human Values*, vol 26, pp43–56

Echols, M. (2001) *Food Safety and the WTO: The Interplay of Culture, Science and Technology*, Kluwer, The Hague

French, D. (2002) 'The role of the state and international organizations in reconciling sustainable development and globalization', *International Environmental Agreements: Politics, Law and Economics*, vol 2, pp135–150

Fulponi, L. (2006) 'Private voluntary standards in the food system: The perspective of major food retailers in OECD countries', *Food Policy*, vol 31, pp1–13

Gawron, J. and Theuvsen, L. (2009) 'Certification schemes in the European agri-food sector: Overview and opportunities for Central and Eastern Europe', *Outlook on Agriculture*, vol 38, no 1, pp9–14

Gille, Z. (2006) 'Detached flows or grounded place-making projects?', in G. Spaargaren,

A. Mol and F. Buttel (eds) *Governing Environmental Flows: Global Challenges to Social Theory*, The MIT Press, Cambridge, MA, and London, pp137–156

Goverde, H. and Nelissen, N. (2002) 'Networks as a new concept for governance', in P. Driessen and P. Glasbergen (eds) *Greening Society*, Kluwer, Dordrecht, pp27–45

Griffin, K. (2003) 'Economic globlization and institutions of global governance', *Development and Change*, vol 34, no 5, pp789–807

Gulbrandsen, L. (2004) 'Overlapping public and private governance: Can forest certification fill the gaps in the global forest regime?', *Global Environmental Politics*, vol 4, no 2, pp75–99

Hall, R. and Biersteker, T. (2002) 'The emergence of private authority in global governance', in R. Hall and T. Biersteker (eds) *The Emergence of Private Authority in Global Governance*, Cambridge University Press, Cambridge, pp3–22

Hatanaka, M., Bain, C. and Busch, L. (2005) 'Third-party certification in the global agrifood system', *Food Policy*, vol 30, pp354–369

Held, D. (2004) *Global Covenant: The Social Democratic Alternative to the Washington Consensus*, Polity Press, Cambridge

Held, D., McGrew, A., Goldblatt, D. and Perraton, J. (1999) *Global Transformations: Politics, Economics and Culture*, Polity Press, Cambridge

Henson, S., Masakure, O. and Cranfeld, J. (2009) *Do Fresh Produce Exporters in Sub-Saharan Africa Benefit from GlobalGAP Certification?*, University of Guelph, Guelph

Humphrey, J. (2008) *Private Standards, Small Farmers and Donor Policy: EUREPGAP in Kenya*, IDS, Brighton

Karkkainen, B. (2004) 'Post-sovereign environmental governance', *Global Environmental Politics*, vol 4, no 1, pp72–96

Kastner, J. and Pawsey, R. (2002) 'Harmonising sanitary measures and resolving trade disputes through the WTO-SPS framework. Part I: A case study of the US-EU hormone-treated beef dispute', *Food Control*, vol 13, pp49–55

Kjaernes, U., Harvey, M. and Warde, A. (2007) *Trust in Food: A Comparative and Institutional Analysis*, Palgrave MacMillan, Houndmills

Klintman, M. and Boström, M. (2004) 'Framings of science and ideology: Organic food labelling in the US and Sweden', *Environmental Politics*, vol 13, no 3, pp612–634

Krahmann, E. (2003) 'National, regional, and global governance: One phenomenon or many?', *Global Governance*, vol 9, pp323–346

Lang, T. and Heasman, M. (2004) *Food Wars. The Global Battle for Mouths*, Earthscan, London

Lang, T., Barling, D. and Caraher, M. (2009) *Food Policy: Integrating Health, Environment and Society*, Oxford University Press, Oxford

Lee, R. (2009) 'Agri-food governance and expertise: The production of international food standards', *Sociologia Ruralis*, vol 49, no 4, pp415–431

Lemos, M. and Agrawal, A. (2006) 'Environmental governance', *Annual Review of Environment and Resources*, vol 31, pp297–325

Lipschutz, R. and Fogel, C. (2002) '"Regulation for the rest of us?" Global civil society and the privatization of transnational regulation', in R. Hall and T. Biersteker (eds) *The Emergence of Private Authority in Global Governance*, Cambridge University Press, Cambridge, pp115–140

Litfin, K. (2000) 'Environment, wealth, and authority: Global climate change and emerging modes of legitimation', *International Studies Review*, vol 2, no 2, pp119–148

Marsden, T., Lee, R., Flynn, A. and Thankappan, S. (2010) *The New Regulation and Governance of Food: Beyond the Food Crisis?*, Routledge, New York and London

Mol, A. (2001) *Globalization and Environmental Reform: The Ecological Modernization of the Global Economy*, The MIT Press, Cambridge, MA

Mol, A. (2010) 'Environmental authorities and biofuel controversies', *Environmental Politics*, vol 19, no 1, pp61–79

Motaal, D. (1999) 'The agreement on technical barriers to trade, the Committee on Trade and the Environment and ecolabelling', in G. Simpson and W. Chambers (eds) *Trade, Environment and the Millennium*, United Nations University Press, Tokyo, pp223–238

Narlikar, A. (2005) *The World Trade Organization: A Very Short Introduction*, Oxford University Press, Oxford

Oosterveer, P. (2006) 'Environmental governance of global flows: The case of labeling stategies', in G. Spaargaren, A. Mol and F. Buttel (eds) *Governing Environmental Flows: Global Challenges to Social Theory*,The MIT Press, Cambridge, MA, pp267–301

Oosterveer, P. (2007) *Global Governance of Food Production and Consumption: Issues and Challenges*, Edward Elgar, Cheltenham and Northampton

Oosterveer, P. (2008) 'Governing global fish provisioning: Ownership and management of marine resources', *Ocean & Coastal Management*, vol 51, pp797–805

Oosterveer, P. and Spaargaren, G. (2011) 'Organising consumer involvement in the greening of global food flows: The role of environmental NGOs in the case of marine fish', *Environmental Politics*, vol 20, no 1, pp97–114

Oosterveer, P., Guivant, J. and Spaargaren, G. (2007) 'Shopping for green food in globalizing supermarkets: Sustainability at the consumption junction', in J. Pretty, A. Ball, T. Benton, J. Guivant, D. Lee, D. Orr, M. Pfeffer and H. Ward (eds) *The Handbook of Environment and Society*, Sage Publications, London, pp411–428

Otsuki, T., Wilson, J. and Sewadeh, M. (2001) 'Saving two in a billion: Quantifying the trade effect of European food safety standards on African exports', *Food Policy*, vol 26, pp495–514

Paterson, M. (1999) 'Overview: Interpreting trends in global environmental governance', *International Affairs*, vol 75, no 4, pp793–802

Polanyi, K. (1944) *The Great Transformation*, Beacon Press, Boston

Potter, C. and Burney, J. (2002) 'Agricultural multifunctionality in the WTO: Legitimate non-trade concern or disguised protectionism', *Journal of Rural Studies*, vol 18, pp35–47

Raynolds, L. (2000) 'Re-embedding global agriculture: The international organic and fair trade movement', *Agriculture and Human Values*, vol 17, pp297–309

Rosenau, J. (2007) 'Governing the ungovernable: The challenge of a global disaggregation of authority', *Regulation & Governance*, vol 1, pp88–97

Sassen, S. (2006) *Territory, Authority, Rights: From Medieval to Global Assemblages*, Princeton University Press, Princeton and Oxford

Schaeffer, R. (1995) 'Free trade agreements: Their impact on agriculture and the environment', in P. McMichael (ed) *Food and Agrarian Orders in the World-Economy*, Praeger Publishers, Westport, pp255–275

Skelcher, C. (2009) 'Fishing in muddy waters: Principals, agents, and democratic governance in Europe', *Journal of Public Administration Research and Theory*, vol 20, ppi161–i175

Spaargaren, G. and Mol, A. (2008) 'Greening global consumption: Redefining politics and authority', *Global Environmental Change*, vol 18, no 3, pp350–359

Speth, J. (2002) 'The global environmental agenda: Origins and prospects', in D. Esty and M. Ivanova (eds) *Global Environmental Governance: Opinions & Opportunities*, Yale School of Forestry and Environmental Studies, New Haven, pp11–30

Stripple, J. (2006) 'Editorial. Rules for the environment: Reconsidering authority in global environmental governance', *European Environment*, vol 16, pp259–264

Tansey, G. and Worsley, T. (1995) *The Food System. A Guide*, Earthscan, London

Treib, O., Bähr, H. and Falkner, G. (2007) 'Modes of governance: Towards a conceptual clarification', *Journal of European Public Policy*, vol 14, no 1, pp1–20

Unnevehr, L. and Jensen, H. (1999) 'The economic implications of using HACCP as a food regulatory standard', *Food Policy*, vol 24, pp625–635

Van der Grijp, N. and Hond, F. (1999) *Green Supply Chain Initiatives in the European Food and Retailing Sector*, IVM, Amsterdam

Van Kersbergen, K. and Van Waarden, F. (2004) '"Governance" as a bridge between disciplines: Cross-disciplinary inspiration regarding shifts in governance and problems of governability, accountability and legitimacy', *European Journal of Political Research*, vol 43, pp142–171

Van Tatenhove, J., Arts, B. and Leroy, P. (eds) (2000) *Political Modernisation and the Environment: The Renewal of Environmental Policy Arrangements*, Kluwer, Dordrecht

VWA (Verner Wheelock Associates) (2005) *Report of Nitrate Monitoring Results Concerning Regulation EU 466/2001: The Netherlands 2004*, VWA, The Hague

Wapner, P. (1997) 'Governance in global civil society', in O. Young (ed) *Global Governance: Drawing Insights from the Environmental Experience*, The MIT Press, Cambridge, MA, pp65–84

Wapner, P. (1998) 'Reorienting state sovereignty: Rights and responsibilities in the environmental age', in K. Litfin (ed) *The Greening of Sovereignty in World Politics*, The MIT Press, Cambridge, MA, pp275–297

Young, O. (2000) *Global Governance: Drawing Insights from Environmental Experience*, The MIT Press, Cambridge, MA

Zaccaï, E. (ed) (2007) *Sustainable Consumption, Ecology and Fair Trade*, Routledge, Abingdon

Section II

# Case Studies

# 5

# Food Provisioning and Climate Change

This chapter aims to:

- present background information on the relative contribution from food provision to global climate change;
- illustrate the consequences of global warming for food production in Africa;
- discuss different strategies to reduce the impact of food production and consumption on global warming.

## Introduction

Climate change is one of the most urgent global environmental problems. Rising temperatures and sea levels and increasingly volatile weather conditions may threaten the livelihoods of many people. At the same time, global warming is, at least in part, the consequence of human behaviour. Scientific experts organized in the Intergovernmental Panel on Climate Change (IPCC) concluded in their latest report that 'most of the observed increase in global average temperatures since the mid-20th century is very likely due to the observed increase in anthropogenic [GHG] concentrations' (IPCC, 2007, p39).

Food production, processing, trade and consumption are considered important contributors to global warming as emitters of GHGs. At the same time, climate change also has direct impacts on the world's capacity to produce food. Particularly in fragile areas, drought and adverse weather conditions may endanger the livelihoods of people who depend on self-provisioning through small-scale agriculture. Global warming will add to already-severe problems of food insecurity by exacerbating problems in food production and trade, stability of supplies and food access (Early, 2009). Hence, action needs to be taken to prevent further climate change (*mitigation*), including reducing emissions of GHGs from food provisioning. Reducing the impacts from climate change by changing farming practices (*adaptation*) also seems necessary to maintain the production capacity of fragile regions (Meinke and Stone, 2005).

There remains much debate, however, on which concrete measures should be taken. This debate on strategies relates, at least in part, to competing views on the relationship between global warming and globalization. Some consider globalization to be an important cause of climate change, as the growth of global food markets encourages the formation of a food market based on worldwide transport (air, ship and road), and more local and regional ways of food provision are needed to reverse this trend. Others, however, suggest that international trade contributes to improved energy efficiency in food production, while being vital for the food security of people living under fragile conditions. Different strategies emerge from these competing views when they are translated into action, each with their particular consequences for food security, food consumption practices and production conditions in both developed and developing countries. It is therefore important to review the relationship between food provisioning and climate change, as well as the proposed strategies that emanate from this relationship.

This chapter first reviews the impact of food provisioning on climate change and presents some key indicators and measurement tools. This is followed by an overview of the different strategies, management tools and governance arrangements intended to mitigate the impact of food provisioning on climate change. Perspectives on adaptation to climate change are also briefly addressed. These strategies are then compared to determine critical issues, which are examined as well. Finally, the main conclusions are elucidated.

## Background Information on the Relationship between Food Provisioning and Climate Change

Global warming has already led to a rise in temperatures and is expected to lead to even further rises in the future. The IPCC expects that without important changes in human behaviour, the temperature may rise by as much as 4°C by the end of this century, with even a 2°C increase resulting in irreversible climate change. Global warming caused by the continued emissions of GHGs (see Box 5.1) means higher temperatures, rising sea levels and increasingly volatile weather conditions.

### Box 5.1 Greenhouse gases

Among gases contributing to climate change, carbon dioxide ($CO_2$) is the most well known, but others such as methane ($CH_4$) and nitrous oxide ($N_2O$) are much more important in the case of agriculture. Each gas has its own global warming potential, such that methane is considered to contain 23 times and nitrous oxide 296 times more global warming potential than carbon dioxide. GHG emissions are measured in $CO_2$-equivalents ($CO_2e$), where methane, nitrous oxide and other contributors to climate change are multiplied by their carbon dioxide equivalents to produce a single, standardized sum.

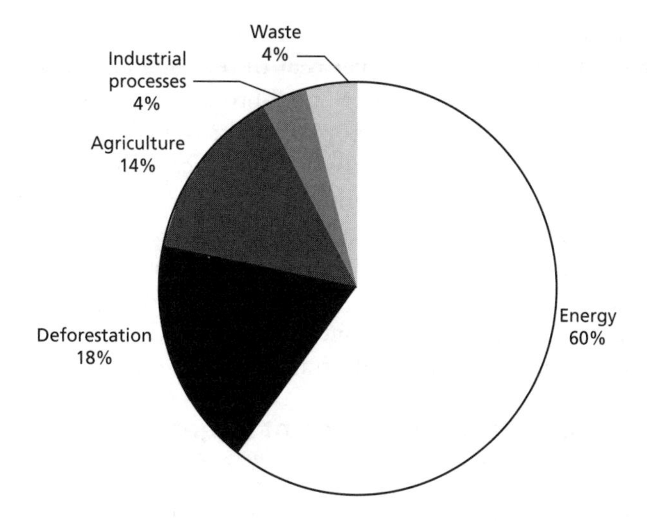


**Figure 5.1** *Sources of GHG emissions*

*Source:* UNCTAD (2009)

## Box 5.2 Taking stock

A study in the southeast UK showed that food and drinks were responsible for over 25 per cent of the total ecological footprint for all activities. A further breakdown was made, and this proved that meat and dairy consumption was responsible for 66 per cent, cereals and other plant-based food for 23 per cent, all drinks 6 per cent and packaging 4.5 per cent of the climate-change impact.

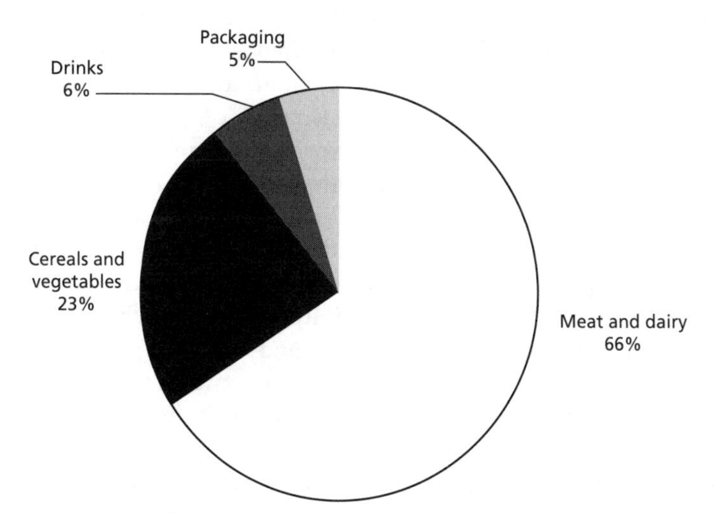


**Figure 5.2** *Relative contributions from different food-related activities to GHG emissions (UK)*

*Source:* Barrett et al (2003)

Agriculture is an important contributor to global warming because it is considered responsible for about 13.5 per cent of the total annual generation of GHG emissions (IPCC, 2007). Primary food-production activities generate about 47 per cent of all global anthropogenic emissions of methane (mainly from animal flatulence and off-gassing from manure, rice fields and peat soils), and 58 per cent of global emissions of nitrous oxide (from synthetic nitrogen fertilizer and tilled soils) (ICTSD-IPC, 2009). Direct $CO_2$ emissions resulting from fossil energy use in transport, cooling, processing and machinery further increase this impact from food-related activities. Additional stages in food provisioning and transport may also contribute to GHG emissions. A brief exploration of these issues will further illustrate these issues (see Box 5.2).

## Climate-relevant impacts at the different stages in the food-supply chain

Each stage in the food-supply chain, 'from farm to fork' and beyond (see Figure 5.3), generates GHG emissions, exacerbating climate change. Primary agricultural production is a recognized contributor to global warming and included as a separate category in global and national overviews (Kramer et al, 1999; Tukker and Jansen, 2006; IPCC, 2007; Carlsson-Kanyama and Gonzalez, 2009). These calculations mostly entail the impacts directly related to farming, but it may be useful to include other factors as well, such as:

- *Land-use change.* Land-use change (converting natural forests and peat land to farm lands) is rarely included, though it may have a substantial impact on the selection of mitigation strategies. For instance, the growing marketing, processing and, ultimately, demand for beef are met by an expansion of managed grazing lands, leading to land-use/land-cover change, especially in tropical and subtropical regions. Destabilization of the climate figures prominently among these disruptive effects (McAlpine et al, 2009). But also, *shifting to organic agriculture* is expected to require more land because of the lower productivity associated with that type of farming.
- *Energy use.* One impact that generally is included when calculating farming-induced climate-change activities is the use of energy for production. Chemical fertilizers are used to increase crop yields, but manufacturing them relies on fossil fuels. Reducing emissions by lowering the chemical intensiveness of inputs may lead to stagnant yields and increased pressure on forests and grasslands (Early, 2009). Other relevant activities are the application of pesticides and the use of machinery operating on the basis of fossil fuels. Next, one should include the energy needed for heating greenhouses and polytunnels when growing vegetables under adverse climatic conditions, and for transporting feed for intensive animal husbandry.
- *Transport and cold storage.* After harvesting the primary products, fossil fuels are needed for transporting food. In most climate-change calculations, this contribution is often classed under the general heading of fossil-fuel use

for transport. The impact of this stage is highly dependent on the distance and means of transport (ship, train, truck, plane) used. Energy is also used in the food-processing stage, with the quantities varying widely among different final products, for instance, producing fresh salad mixes, frozen peas or ready-to-bake frozen pizzas. The next phase of transfer from factory to wholesaler, and ultimately to retailers, demands additional fossil fuels for transport and (cold) storage.

- *Consumer practices*. Rarely is the role of households, the final consumer, included in climate-change studies; however, research shows that their contributions may be substantial. For instance, Foster et al (2006, p143) observe that 'the environmental impacts of car-based shopping (and subsequent home cooking for some foods) are greater than those of transport within the distribution system itself'. And Van Hauwermeiren et al (2007) conclude, on the basis of an empirical study, that energy use and carbon dioxide emissions are almost always higher in local food-supply systems (full summer, inland production) than in conventional food systems. They explain this by pointing to the low energy efficiency when consumers transport food from the farm or the shop to home. The net effect of this activity depends on the means by which food is transported from the retailer to home and how efficiently this is done. Household food storage (freezing and refrigeration) and preparation (food processing, cooking, microwaving, etc.) put substantial claims on energy as well. The amount of food wasted may be surprising. For instance, Carlsson-Kanyama and Gonzalez (2009) report that 20 to 30 per cent of all food is wasted. Food waste makes a considerable quantity of captured GHGs useless, and minimizing this waste would be an effective contribution to reducing climate change.

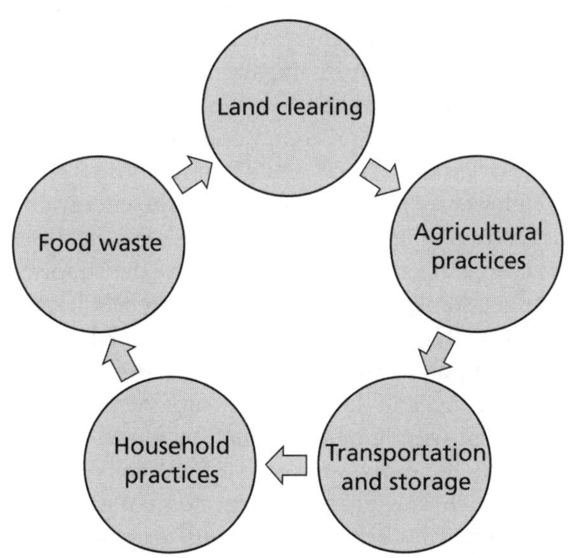


**Figure 5.3** *Model of life-cycle analysis of food provisioning*

## Measuring the contributions from food provisioning to climate change

Identifying possible strategies to reduce the climate impact from food provisioning requires tools to measure this impact. Over time, various measurement indicators have been introduced (OECD, 2000), but most focus on primary agricultural production and pay little attention to the rest of the supply chain. Two approaches that are more integrative are discussed here: LCA and the calculation of energy used and GHG produced.

LCA examines environmental impacts throughout the course of a product's life cycle: from raw material acquisition through production, to use and disposal (see Chapter 3). Foster et al (2006) use the LCA methodology on a small number of products that are representative of all food consumption in the UK, to inform government agencies on how to reduce the environmental impacts of food production and consumption. The authors conclude that the environmental impacts of organic agriculture are often lower than for conventionally grown food. However, 'this is not true for all foods and appears seldom to be true for all classes of environmental impact' (Foster et al, 2006, p141). They found only weak evidence for a positive environmental impact from local preferences in food supply and consumption when all food types are taken into consideration. Lastly, the authors found that food packaging can have high environmental impacts, depending on how it is recycled or disposed after use. Such full LCAs are very helpful, but the problem is that they are usually relatively costly and time consuming to apply (Bolwig et al, 2008).

Carlsson-Kanyama (1998) developed an often-used indicator to measure the environmental impact of food production on the basis of energy used. This indicator calculates the GHG emissions (in $CO_2e$) and energy consumption (in megajoules per kilogram (MJ/kg) product) of food products. The system boundaries are put at the end of the production chain before the product reaches the consumer to facilitate comparisons; hence, the indicator does not include consumer impacts. When this measurement tool is applied, the findings show emissions of GHG and energy consumption differ greatly among various food items. For instance, pork emits 9 times more GHG than dry peas, and rice emits 38 times more than potatoes. And, the energy consumption for tomatoes is almost 15 times higher than for carrots. Among different products, there are large variations regarding which phase of the supply chain contributes most. Surprisingly, transportation proved not to be a large contributor to the total emissions of GHG, although it often accounted for a major part of the total energy consumption. The analysis of GHG showed that such stages as storage, animal rearing and farming were major contributors, but this does not become evident from an energy analysis alone. The study therefore observed that 'conclusions about sustainable food consumption based on a single functional unit such as weight or energy content of the food itself must be drawn carefully, otherwise recommendations may be skewed' (Carlsson-Kanyama, 1998, p291).

Although these tools, which apply scientific data to judge the sustainability of a particular food, provide interesting information, their contribution to effecting change seems limited. The absence of a transition perspective results in little guidance on how to increase the sustainability of an existing practice. A complicating factor is the absence of consumer involvement in applying these tools; the results are sometimes counter-intuitive for consumers.

## Expected Climate-change Impacts in Developing Countries

Climate change may impact agriculture everywhere. In some regions, impacts may actually be positive. For instance, at mid to high latitudes with local mean-temperature rises of up to 1–3°C, crop productivity is projected to increase slightly depending on the crop, and then decrease beyond that in some regions (IPCC, 2007). In other regions, however, the impacts are expected to be primarily negative, and sub-Saharan Africa deserves particular attention, as:

> *climate is projected to change strongly in sub-Saharan Africa, with annual average temperature increases there between 1.8 and 4.8°C and annual changes in rainfall ranging between –12 and +25 per cent (seasonal changes range from –43 to +38 per cent) by 2100. Severe impacts on agricultural production throughout the continent are therefore expected.* (Müller, 2009, pp29–30)

Higher temperatures will immediately increase stress on crops; as no crop in Africa is constrained by low temperatures, there will be only disadvantages. Higher temperatures will speed up evaporation from plants and soils. Research suggests that rain-fed agriculture will reduce its production by up to 50 per cent between 2000 and 2020 because of the reduced growing season and increased heat stress on plants (Toulmin, 2009). In its 2007 report, the IPCC takes an inventory of the most prominent expected impacts from climate change in Africa and Asia. At lower latitudes, especially in seasonally dry and tropical regions, crop productivity is projected to decrease for even small local temperature increases (1–2°C), which would increase the risk of hunger. In Africa, the IPCC expects that by 2020 between 75 and 250 million people will be made vulnerable because of increased water stress. Agricultural production, including access to food, in many African countries is projected to be severely compromised. This would further adversely affect food security and exacerbate malnutrition. Rainfall patterns would probably shift, with most decreases in West, Central and Southern Africa and increases in East Africa. Extreme weather events will become more frequent because of the greater amount of moisture in the air. Other potential impacts are changing distribution patterns in pests and diseases and impacts on crop pollinators: bats, bees and moths. Livestock may have to move to other areas, with increased exposure to infections and diseases as a consequence, while higher temperatures and greater moisture intensify these effects even further. A small positive contribution would come from the increased $CO_2$ levels in the atmosphere, as they would provoke greater photosynthesis.

Mitigation measures in sub-Saharan Africa could be the reduction of deforestation, technological development to reduce emissions and increased carbon sequestration above and below ground. Agricultural adaptation measures in sub-Saharan Africa are possible, such as changes in agricultural practices and relocation of agricultural production. This will be very difficult in low latitudes, however, if temperatures increase by more than 3°C. The best measures are policies targeted at improving agricultural productivity, rural physical and institutional infrastructure and poverty alleviation in rural areas, because they not only contribute to food security and poverty alleviation, but also increase the agricultural sector's resilience (ICTSD-IPC, 2009). The cost of adaptation could amount to at least 5 to 10 per cent of these countries' gross domestic product (GDP). Adaptation needs to combine a focus on building more resilient local food and water systems, national food-security plans and international management of key resources, such as river basins (Toulmin, 2009). Most agriculture (95 per cent) in sub-Saharan Africa is rain-fed and does not use irrigation systems, making it very vulnerable. African farming systems are diverse and, raising livestock is often central (30 per cent of the value of agricultural production). One approach is to strengthen already-existing indigenous coping strategies, such as diversifying crops and animals, sending family members out of the countryside to cities and moving to higher rainfall areas. Adaptation measures for livestock include continued mobility to avoid overuse of more limited resources. Overall, however, observers fear that competition for these dwindling resources in sub-Saharan Africa will increase.

Unfortunately, Africa will be severely hit by climate change, while the continent has contributed very little to its cause: global warming (1 tonne $CO_2$ per capita compared with 4.3t $CO_2$ as the global average, with the US at 19.9t $CO_2$, the EU 6.9t $CO_2$ and China at 3.2t $CO_2$ for the year 2007) (Toulmin, 2009). At the same time, because of its weak economic power and internal divisions, the continent has little influence on decisions taken at the global level to counter the impacts.

In Asia, the main concerns are decreasing freshwater availability for the growing population in many parts of the continent and river flooding in the heavily populated mega-delta regions in South, East and Southeast Asia (Kameyama et al, 2010). Of particular concern is the risk that melting glaciers in the Himalayas will increase the volatility in many of the large Asian river basins and thereby put irrigation farming under pressure. At the same time, the expected sea-level rise may further increase the vulnerability of livelihoods and food production in the low-lying coastal zones, such as in the case of Bangladesh.

## Policy Tools to Address Agriculture, Food and Climate Change

Multiple strategies have been generated to address climate problems and their connections with farming and food provision presented above. Among these strategies we find international negotiations and multilateral environmental agreements, such as that on Kyoto. But also, various governments have developed particular national policies through legal regulations and voluntary agreements, while companies and consumers are contributing in their own way. Below we discuss the most important intervention strategies.

### International negotiations

The United Nations Framework Convention on Climate Change (UNFCCC), adopted in 1992, was the first multilateral agreement on climate change and set the goal of reducing GHG emissions. New information on the challenges that climate change posed to social, economic and environmental systems and the range and severity of projected impacts became available through the IPCC's Second Assessment Report in 1995, which provided a new sense of urgency and helped inform the negotiation of the Berlin Mandate, which led eventually to the Kyoto Protocol. The Conference of the Parties (COP), the Convention's supreme body responsible for keeping the implementation of the UNFCCC under review and adopting related legal instruments, adopted the Kyoto Protocol to the UNFCCC in 1997. The principle of equity and common but differentiated responsibilities and respective capabilities was made effective in the Kyoto Protocol by requesting industrialized countries to take the lead through quantified emission limitations and reduction objectives. Thus, the Kyoto Protocol imposed a binding commitment on developed countries to reduce their $CO_2$ emissions by at least 5 per cent – in aggregate – by the period 2008–2012 as compared to 1990 levels. Negotiations under way now take place under both the UNFCCC and the Kyoto Protocol, including those on a new and more aggressive emissions reductions commitment, and were meant to conclude at the Copenhagen summit in December 2009. The meeting in Copenhagen failed to take decisive new steps, however, which resulted in an opening up of the discussion framework, and most observers fear that it will take several years before a new momentum is built to agree on a treaty that will replace the Kyoto protocol.

During the Copenhagen meeting, agriculture was discussed as requiring a 'sectoral approach', and two strategies were debated but not agreed upon, because the meeting as a whole collapsed. The suggested strategies were either to aim for an agreement on sector-specific mitigation efforts through various means, or to promote and cooperate on research, development, application and diffusion of technologies, practices and processes to enhance mitigation. While the impacts of climate change on food security were not debated, those on international trade were discussed. On these impacts, the negotiating countries agreed that climate mitigation or adaptation activities should not result in barriers to or distortion of the international trade system of agricultural products. This means they argued against introducing sectoral targets, carbon-labelling schemes (see below), carbon 'footprinting', border-tax measures or other national approaches to climate mitigation. The negotiating governments argued that the principles of the WTO agreement should be upheld and that no protectionism under the disguise of climate mitigation should be allowed.

### *The EU*

As a supranational governance institution, the EU deserves special attention. There are many specifically food-related policies driven by EU directives aimed at the agricultural and land-use sector. None of these policies is specifically intended to address climate change, although many will have an indirect influence. Most prominent is the Common Agricultural Policy (CAP) that, since its major reform in 2003, is shifting the balance of agricultural support in Europe away from production-oriented agricultural subsidies to area-based payments. Such payments are conditional on 'cross-compliance' with minimum environmental and other standards. The intended outcomes are to provide less of an incentive to produce more and to generate a better focus on creating environmental benefits. The CAP is presently being reviewed to prepare the way for long-term reform. One challenge identified in this process is climate change, but this objective must compete with others, such as further liberalizing product markets, including that for milk, which may lead to rearing more cows and eventually to higher emissions.

With respect to other environmentally oriented regulations, the EU Nitrates Directive (91/676/EC) is of key importance to agricultural sustainability, as it seeks to reduce the impacts of nitrates escaping into water and air. The EU's Integrated Pollution Prevention and Control Directive regulates, and is intended to reduce emissions from, polluting activities, including those associated with intensive pig and poultry farms. The regulations affect raw materials use, waste, slurry and manure management, livestock housing, energy and accident management. Finally, in 2006 the EU adopted a Soil Thematic Strategy aimed at preventing further soil degradation and restoring degraded soils. This led to proposals to establish a Soil Framework Directive that would legally require member states to take measures to protect and improve their soil. At present, however, member states have been unable to reach an agreement on how to implement such a directive (Garnett, 2009).

## National regulations

Building on international agreements, such as the Kyoto Protocol, or under pressure from domestic public opinion, national governments intervene as well. For instance, the UK has incorporated a legal obligation in its national policy to reduce carbon emissions by 80 per cent in 2050, as compared with the 1990 level (the Carbon Reduction Commitment). Governments resort to various political instruments to combat global warming.

An existing instrument is the carbon tax. Carbon taxes are national measures[1] that make producing carbon-intensive products more expensive by including the negative externalities resulting from GHG emissions in the price of a product. Taxation of the carbon content of products has been used in some countries, such as Finland and Sweden, for a number of years. The main principle behind this approach is to ensure that the consumer pays for the environmental consequences resulting from the production and use of a particular product. Imposing such measures unilaterally, however, may make domestic production less competitive than imports and consequently cause such production to move to other countries without comparable mitigation measures. To counter such leakage, countries may seek to impose border tax adjustments, but are faced with the problem that such measures may run counter to WTO obligations (Early, 2009). An alternative measure is based on introducing tradable carbon permits, which may offer reduced costs and greater levels of compliance, encourage the adoption of cleaner technologies and help finance abatement. Allied with tradable carbon permits is an offset policy whereby GHG emissions can be offset by some ameliorating action elsewhere in the economy. Other options for national governments are granting subsidies for reducing energy use and increasing alternative energy production. Public authorities may also require standardized information to be available on (food) products to inform consumers about the global-warming impacts resulting from their buying choices. Another direct governmental intervention would be to include criteria on the climate-change impact of particular food products in public procurement policies. A recent proposal is to hand each citizen a personal carbon allowance that they can use according to their own priorities, but that strictly limits the total climate-change impacts that they are allowed to generate (Fawcett and Parag, 2010).

## Voluntary instruments

Besides governments, NGOs and private companies also have introduced climate-governance tools to encourage changes in GHG-related practices. These instruments have a voluntary character and are based on providing information to producers, processors, retailers and consumers, enabling them to (re-)consider their daily practices in light of climate-change impacts. Among them one can find production guidelines ('best practices'), subsidies to make environmentally friendly activities more attractive and facilitate access to technologies that have reduced impacts on global warming. 'Food miles' were one of the first consumer-oriented concepts to be introduced, and this was later supplemented with the more elaborate 'carbon labelling'.

### *Food miles*

Food miles have emerged since the mid-1990s as a clear discursive framework, particularly in the US, with an emphasis on supporting local (especially family-based) agricultural producers (Iles, 2005). Food transported over long distances by air and trucks is associated with high carbon emissions and energy use and is therefore considered to have negative impacts on global warming. After the 2003 foot-and-mouth disease crisis, food miles became an influential heuristic in the UK as well and is visibly included in decisions taken by such supermarket retailers as Sainsbury's, Tesco and Marks & Spencer (Iles, 2005). Food miles are intended to be a simple and objective measurement of the sustainability of food because they measure the environmental effects of transporting foodstuffs from the primary producer, via processing industries, to retail outlets and then to consumers (Pretty et al, 2005). The energy required for this transport is considered a heavy burden for the global climate, and therefore reducing these unnecessary $CO_2$ emissions would be very helpful. When calculated, food miles help make visible the environmental impacts of long-distance transportation compared with locally sourced foods.[2] By the year 2000, for example, the average food item in the US had travelled between 2400 and 3200 kilometres from farm to plate, 25 per cent more than in 1980 (Pirog et al, 2001). If all food were organic and locally sourced or predominantly transported by rail and then transported home by walking/cycling, bus or home delivery, these external costs would fall from 11.8 to 1.8 per cent or less of the average food basket (Pretty et al, 2005). Other considerations that fit the same argument are that local food is tastier because it is fresh and that the local economy is supported, while vulnerability to oil shortages and dependence on whimsical international markets are reduced (Halweil, 2002). In response to these concerns, the UK supermarket chain Tesco decided to put aeroplane symbols on all food products air-freighted to the UK.

Over time, the reference to food miles as a straightforward and effective instrument for reducing climate change has received serious criticism. Commentators question two aspects: first, whether physical distance between food producer and consumer can be considered a good measurement for a food item's GHG emissions, and second, whether the global consequences of such a strategy are morally or politically acceptable. Different case studies have shown that not only the physical distance but also the mode of transport (by air, sea or trucking over land) has a direct effect on the climate impact of food. Moreover, the methods involved in producing food, from the harvesting methods to the type of fertilizer used and the fuel consumed to make the packaging, are relevant carbon variables as well. Environmental scientists in New Zealand calculated, for instance, that lamb from New Zealand shipped to Britain has a carbon footprint just one-fourth that of British lamb – in part because most electricity in New Zealand comes from renewable sources, while the availability of sufficient rain and sun means that the country's extensively used pastures need less fertilizer than those in the UK (Saunders and Barber, 2007). Another UK study indicated that only 2.2 per cent of the total carbon footprint from ruminant meat consumption is transport-related (Minx et al, 2008). Finally, food miles are mostly calculated up to the point of sale, until its arrival on the supermarket shelves, but the impact of transport from point of sale to the buyers' home can be substantial as well (Van Hauwermeiren et al, 2007).

## Box 5.3 The Soil Association and air (food) miles

As mentioned in Box 3.6 in Chapter 3, the Soil Association raised the question in a consultation paper of whether it should address the environmental impact of air freight in its organic standards. The suggestion to ban all imports transported by plane generated a discussion on the impacts for organic farmers in developing countries. Some argued that consumers expect organic products to meet high levels of environmental care, and therefore air transport should be limited. Others pointed out during the consultation process that there is a need to consider carbon emissions for all products and across the whole life cycle. This would mean carbon labelling for every food item, but at the moment this is not possible at affordable costs. At the end of the public debate, the Soil Association concluded that they would not prohibit importing products from developing countries, provided that 'freighted organic food delivers genuine benefits for farmers in developing countries'. Therefore, the Soil Association proposed that such food should meet its own 'ethical trade' standards or equivalent fair trade standards, in addition to the organic standard.

*Source:* Soil Association (2007)

The second aspect of the use of food miles receiving comments is its consequences for food exports from developing countries, because ending food imported via air may threaten the livelihoods of many farmers and workers there. Toulmin (2009) argues that air-freighted fruits and vegetables from Africa account for less than 0.1 per cent of the UK's GHG emissions but provides a livelihood for 1–1.5 million people. The food miles concept is considered one-sided, and aspects of social and economic development should be included to allow for the assessment of trade-offs (MacGregor and Vorley, 2006). The importance of including other dimensions of sustainable development also became clear in a discussion in the organic movement (see Box 5.3).

Another consideration in the debate on food miles is that the relative contributions from transport to the total $CO_2$ impact of various food products differ considerably. While this percentage is relatively high for fresh fruits and vegetables, it is much lower for meat, milk and processed food (MacGregor and Vorley, 2006).

Moreover, most local food production also depends on inputs from natural resources (energy, water, chemicals) that are scarce and often imported from other countries. Hence, local production is not always the most energy-efficient way of producing food. Finally, Weber and Matthews (2008) estimate that the average US household's climate impact related to food is 8.1t $CO_2$e per year, with delivery 'food miles' accounting for around 0.4t $CO_2$e/year and total freight accounting for 0.9t $CO_2$e/year. Shifting a seventh of total calories of red meat/dairy consumption to other protein sources would therefore be as effective as complete localization of food.

These criticisms have led to the search for more reliable and objective methods to calculate the climate-change impact of food items that can guide consumers, retailers and others in their food-purchasing practices. Comprehensive calculation of carbon-emission impacts of particular agricultural products is necessarily complex, but is needed to challenge simplistic, albeit intuitively appealing, concepts such as food miles. Carbon labels may be an important improvement towards achieving this aim.

### *Carbon labels*

Carbon labels are product labels indicating the amount of $CO_2e$ emitted from a product throughout its life cycle. For example, consumers are informed through a label, that a snack-sized bag of Walkers' salt and vinegar potato crisps has a carbon footprint of 75 grams of carbon emissions.[3] Calculating the exact carbon footprint of each product is a daunting task, however, as GHG emissions for each step of the production chain have to be measured, and this becomes very complicated in the case of composed products, such as a pizza or a ready-to-cook meal. An analysis of any food item must include the amount of nitrogen fertilizer used, manure application, processing degree, transportation mode and distance, presence of red meat and storage method. 'The data demands for calculation of the ecological footprint of a product are just a high as for an LCA study. The main difference is in the way the inputs and emissions are aggregated and interpreted' (Bolwig et al, 2008, pp32–33) (see Box 5.4 for an example).

The complexity and costs involved in calculating complete carbon footprints for each food product led to the introduction of simpler labels. Examples are comparative labels that certify that a product is emitting less $CO_2$ than conventional products, sometimes even using colours to indicate the product's performance from a climate-change perspective. There are also several schemes for certifying carbon offsets, which means that the $CO_2e$ emitted during the production process of a food product is compensated by planting trees or another way of reducing GHG emissions. Yet another label certifies that a product lives up to the best practices available within its domain of food production, processing and packaging (Röös et al, 2010).

## Box 5.4 PAS 2050

Publicly Available Specification (PAS) 2050 was developed by the British Standards Institution in collaboration with DEFRA and the Carbon Trust to calculate the GHG emissions from horticultural products. This specification was further elaborated in the Netherlands for calculating $CO_2$ footprints for horticulture on a cradle-to-gate basis for business-to-business information-exchange purposes.

The purpose of this method is to attribute the GHG emissions to the end product. First, the system and its boundaries are defined, then the points of departure for calculating the GHG emissions are identified, and third, the emissions are allocated to the end product (on the basis of economic value).

A computer-based tool has been developed for horticulturalists that allows them to calculate GHG emissions per crop and for each step in the production chain: growth, transport, materials and processing, and soil and fertilization. These emissions are calculated on the basis of (the best available) standard data, and only when information exists.

*Source:* Blonk et al (2009)

All carbon labels are widely debated on several issues. First, there is the technical challenge of how to measure impacts on climate change accurately, particularly when different locally specific technologies are used. Next is the question of how the costs of labelling food products can be reduced, which is particularly relevant for small-scale farmers, especially in developing countries. Another debate concerns the question whether carbon labels allocate responsibilities for environmental impacts correctly and provide adequate tools for action. The information available on the labels should encourage action to reduce the climate impact, but this is not always as simple as expected. Like many eco-labels, carbon labels also are based on the assumption that the general public is not well-enough informed and is in need of objective information from independent, reliable sources and then will adapt its behaviour accordingly. This 'public information deficit' approach ignores that 'facts' are never as unproblematic as suggested within contentious areas of debate. Consumers interpret information within their own frame of reference and have to apply recommendations according to their own lifestyles (Irwin, 2001). Findings generated using a full carbon-footprint methodology that are counter-intuitive may challenge public confidence in such instruments. Therefore, when carbon labels indicate that air-freighted food products have fewer GHG emissions than locally produced food, consumers are encouraged to question the measurement's validity rather than reconsider their buying behaviour. Hence, carbon labels may help inform consumers on the climate-change impact of food products, but changing consumer behaviour in response to this information is more complicated.

## Consumers, food and climate change reduction

As climate change attracts much public interest, it provides a fertile ground for private initiatives from NGOs and consumers. These initiatives suggest different ways to reduce the impact on global warming from the foods people consume. Consuming food in global modernity forces people to rely on abstract systems, scientific expertise and various information systems when making long-distance assessments of the product and of the production methods involved, including their impact on climate. Ambiguity, or balancing trust and risk in food systems, has become an ingrained trait of the contemporary food supply, making many consumers feel insecure about whom and what to trust. Consumer trust in (information about) food also evolves over time and can acquire different shapes depending on various factors. One way forward in this context is the emergence of innovative governance arrangements next to and beyond national governmental regulations. Many of these governance tools are introduced by environmental NGOs, and some of them address global food-supply chains through labels/certification and other means. Consumers and civil-society organizations then become co-drivers of global environmental change. Three ways in which consumers can become involved are as ecological citizens, as political consumers and through their life(style) politics (Spaargaren and Oosterveer, 2010). As citizens, consumers are engaging with public authorities, so when discussing voluntary arrangements, we limit ourselves to political consumerism and food consumer life(style) politics (see also Chapter 11).

### *Political consumerism*

Political consumerism recognizes the consumers' participation in environmental change through politics outside the traditional political arenas ('subpolitics'; Beck, 1992) by boycotts and buycotts (Micheletti, 2003). Political consumerism refers to more voluntary, ad-hoc organized, mostly civil society-based forms of (environmental) politics. The concept brings together all the political forms that connect environmental activities of up-stream economic actors in production–consumption chains and networks more directly and visibly with the interests and activities of citizen-consumers at the lower end of these chains and networks, and vice versa. Political consumers may refuse to buy (boycott) food products with high carbon footprints and be willing to buy (buycott) food with smaller carbon footprints. When allowing consumers to use their buying power, they need to be supplied with reliable information on the climate impact of food products from trustworthy organizations, i.e. devices to bridge social and geographical distances (Oosterveer and Spaargaren, 2011). Consumers can become co-drivers of global environmental change, and their power can be used for political purposes through diverse networks covering actors and interests from civil society, the market and the state. At present, civil-society organizations seem more capable of establishing trust relationships with consumers than state- and market-based organizations.

### *Lifestyle politics*

Life(style) politics are politics of the life-world (Giddens, 1991) and as such, are directly connected to the morals and choices that are implied in ordinary consumption routines. For instance, reduction of food wastes may make a considerable contribution to reducing the climate impact from food provision. Also, the choice of food and diet can have a large impact on the associated GHG emissions. In an overview study, Carlsson-Kanyama and Gonzalez conclude that 'meals with similar nutritional value had a difference in GHG emissions of up to a factor of 4' (2009, p1704S). Furthermore, they observe that environmentally friendly meals are not contradictory to healthy meals, as 'a Mediterranean diet, which consists mainly of plant-origin food but not excluding a small proportion of meat and other animal products, is closer to public health recommendations issued by the WHO and has lower environmental effect than the current average US diet' (Carlsson-Kanyama and Gonzalez, 2009, p1704S). An analysis of the GHG impact from any food item must include the degree of processing, transportation mode and distances, presence of red meat, amount of nitrogen fertilizer used, manure application and storage method.

## Impacts of $CO_2$-Reduction Strategies on Developing Countries

Climate change is a global problem, but measures to reduce its impact may not have a similar impact everywhere. As mentioned above, the impacts on developing countries deserve particular attention. Sub-Saharan Africa is especially vulnerable to global warming, as many regions are already stressed by the lack of water and forest coverage, and climate change will probably exacerbate this. The suggested ways to reduce the climate change impact from food discussed here, however, may further intensify these and other already-existing problems.

Incorporating carbon footprints as a criterion in international food trade may have both negative and positive impacts on trade from developing countries (Brenton et al, 2009). Among the negative impacts, we include the lack or even absence of primary data on GHG emissions, which forces reliance on secondary (less representative) data parameters that may not be adapted to local contexts. Data on the carbon efficiency of production technologies that are specific to particular local production systems are often lacking. Next, land-use changes belong to the more important contributors to global warming, and most of these conversions occur in developing countries. Finally, other negative impacts concern the costs involved in carbon labelling and certification, which may be insurmountably high, especially for smallholder farmers. Among the positive impacts, it may be expected that developing countries are in an advantageous position because they generally use more renewable energy and less carbon-intensive technologies. An interesting side effect could be the relocation of processing facilities to developing countries because transporting final products would require less energy than would be needed for the raw food items.

An important setback is that developing countries and small-scale farmers have little or no representation in the debates on the emerging standards and labelling initiatives, as they are mostly initiated by developed countries retailers and NGOs. It would be commendable to explicitly assess the social and economic impacts, as well as the secondary environmental impacts, of evolving carbon-labelling standards on farmers in developing countries.

## Conclusions

This chapter analysed the relationship between global warming and food provisioning and provided evidence for the close connections between them. Food production, processing and trade have a substantial impact on climate change, while food production may be threatened by global warming, especially in fragile areas, such as in sub-Saharan Africa. Reversing this trend and adapting to the changes already taking place constitute challenges to present-day agri-food-supply systems. In this chapter, we discussed some of the options available for reducing the climate-change impact from food provisioning and reviewed options for product labelling through food miles and carbon footprints.

Although the food-miles approach seems attractive, because it is a simple and straightforward instrument, its unwanted negative consequences suggest that it might be better to replace it with a more refined approach, such as carbon footprints. Nevertheless, this tool also contains several controversial elements. Its measurement is based on a systems approach that necessarily includes truncations (land-use change is often not included, nor are household/consumers' contributions in transporting, storing, preparing and wasting food). In some instances, the processing stage is not included at all. Secondary data often have to be used instead of primary data, although production practices may vary considerably. Finally, the use of standard data in labelling schemes does not reward improvements by individual producers and thus may keep them from applying innovations.

Another discussion concerns the usefulness of carbon labelling as a driver for change: what it can contribute is not clear, except for appealing to consumers to change their eating habits. Carbon labelling does not provide clear incentives for technological change in the production process or for reorganizing the supply chain. The approach is very much based on expert knowledge and is oriented towards agri-food companies and governmental authorities, while consumers are included only in a rather passive manner, as recipients of information. In this way, consumers are understood as atomistic actors whose buying behaviour will change when the correct information is provided. However, food consumption is more than the incorporation of carbon embedded in food products, but part of wider social practices. The ownership of carbon-footprint labelling is often unclear: who should act, as well as why and how, remains mostly undetermined, especially in the context of global food provisioning where national authorities have only limited control. International consensus is lacking regarding the validity of each indicator in measuring the environmental situation as a whole. Often, implicitly or explicitly, governments are regarded as the authority responsible for implementing the results of studies developing or applying various sustainability indicators of food. Carbon labels may offer useful indications for the climate-related performance of agri-food-supply chains, but they lack attention to other sustainability problems. Public concerns such as biodiversity, landscape protection and animal welfare are not addressed, although they, too, deserve attention from regulatory agencies.

Finally, many proposals on carbon labels and other food-related climate-change initiatives lack adequate consideration of their impact on a global scale. Effective introduction of climate-change impact may result in unexpected and possibly unwanted consequences, such as radical shifts in the geographical distribution of food production and processing. Bringing these consequences into the debate may lead to adaptations in the approach, such as considering the argument from African farmers that they are not responsible for climate problems in the first place, so they do not want to become the primary victim of policies combating the problem.

*Take-home lessons*

- In all phases of the food-supply chain, potentially important impacts on the global climate may be identified.
- People living in developing countries are more vulnerable to the impacts of climate change on food provision, although they are less responsible for its causes.
- Different governmental and private-governance instruments are available to address agri-food's impact on climate change, but many face challenges in terms of the required data and the possibility of unwanted side effects.

## Notes

1 Carbon taxes could also be initiated on an international level, but agreement on such measures still seems far away. Nevertheless, governments may impose them on imported products, but the resulting consequences for international trade may be problematic.

2 All farm externalities, domestic road transport of foodstuffs, government subsidies and consumers' transport of conventional food from the retail outlet to home, result in external costs of 11.8 per cent of the food basket (Pretty et al, 2005).

3 To compare this figure: a full jumbo jet flying from Frankfurt to New York emits 713,000 grams of carbon emissions per passenger.

## Further Reading

Carlsson-Kanyama, A. and Gonzalez, A. (2009) 'Potential contributions of food consumption patterns to climate change', *American Journal of Clinical Nutrition*, vol 89, no 5, pp1704S–1709S: a short but insightful article on the relevance of dietary choices for climate change.

Ingram, J., Ericksen, P. and Liverman, D. (eds) (2010) *Food Security and Global Environmental Change*, Earthscan, London and Washington, DC: presents methodologies, impacts and policies in a comprehensive overview.

Rosegrant, M., Ewing, M., Yohe, G., Burton, I., Huq, S. and Valmonte-Santos, R. (2008) *Climate Change and Agriculture; Threats and Opportunities*, GTZ, Eschborn: illustrates concrete steps to address climate change.

Toulmin, C. (2009) *Climate change in Africa*, Zed Books, London: puts the global climate change debate in a wider framework.

## References

Barrett, J., Cherrett, N., Hutchinson, N., Jones, A., Ravetz, J., Vallack, H. and Wiedmann, T. (2003) *A Material Flow Analysis and Ecological Footprint of the South East*, SEEDA, EcoSys, SEI, Guildford

Beck, U. (1992) *Risk Society: Towards a New Modernity*, Sage, London

Blonk, H., Ponsioen, T. and Scholten, J. (2009) *$CO_2$-voetafdruk: rekenmethode voor tuinbouwketens*, Blonk Milieuadvies, The Hague

Bolwig, S., Ponte, S., du Toit, A., Riisgaard, L. and Halberg, N. (2008) *Integrating Poverty, Gender and Environmental Concerns into Value Chain Analysis: A Conceptual Framework and Lessons for Action Research*, Danish Institute for International Studies, Copenhagen

Brenton, P., Edwards-Jones, G. and Jensen, M. (2009) 'Carbon labelling and low-income country exports: A review of the development issues', *Development Policy Review*, vol 27, no 3, pp243–267

Carlsson-Kanyama, A. (1998) 'Climate change and dietary choices: How can emissions of greenhouse gases from food consumption be reduced?', *Food Policy*, vol 23, no 3/4, pp277–293

Carlsson-Kanyama, A. and Gonzalez, A. (2009) 'Potential contributions of food consumption patterns to climate change', *American Journal of Clinical Nutrition*, vol 89, no 5, pp1704S–1709S

Early, J. (2009) 'Climate change, agriculture and international trade: Potential conflicts and opportunities', *Bridges Trade BioRes Review*, vol 3, no 3, pp8–10

Fawcett, T. and Parag, Y. (2010) 'An introduction to personal carbon trading', *Climate Policy*, vol 10, pp329–338

Foster, C., Green, K., Bleda, M., Dewick, P., Evans, B., Flynn, A. and Mylan J. (2006) *Environmental Impacts of Food Production and Consumption: A report to the Department for Environment, Food and Rural Affairs*, Manchester Business School and DEFRA, London

Garnett, T. (2009) 'Livestock-related greenhouse gas emissions: Impacts and options for policy makers', *Environmental Science & Policy*, vol 12, pp491–503

Giddens, A. (1991) *Modernity and Self-Identity: Self and Society in Late Modern Age*, Polity Press, Cambridge

Halweil, B. (2002) *Home Grown: The Case for Local Food in a Global market*, World-watch Institute, Washington, DC

ICTSD-IPC (International Centre for Trade and Sustainable Development–International Policy Council) (2009) *International Climate Change Negotiations and Agriculture*, ICTSD and IPC, Geneva

Iles, A. (2005) 'Learning in sustainable agriculture: Food miles and missing objects', *Environmental Values*, vol 14, pp163–183

IPCC (Intergovernmental Panel on Climate Change) (2007) *Climate Change 2007: Synthesis Report*, IPCC, Geneva

Irwin, A. (2001) 'Constructing the scientific citizen: Science and democracy in the biosciences', *Public Understanding of Science*, vol 10, pp1–18

Kameyama, Y., Sari, A., Soejachmoen, M. and Kanie, N. (eds) (2010) *Climate Change in Asia: Perspectives on the Future Climate Regime*, United Nations University Press, Tokyo

Kramer, K., Moll, H., Nonhebel, S. and Wilting, H. (1999) 'Greenhouse gas emissions related to Dutch food consumption', *Energy Policy*, vol 27, no 4, pp203–216

MacGregor, J. and Vorley, B. (2006) 'Fair Miles? The concept of "food miles" through a sustainable development lens', *Sustainable Development Opinion*, IIED, London

McAlpine, C., Etter, A., Fearnside, P., Seabrook, L. and Laurance, W. (2009) 'Increasing world consumption of beef as a driver of regional and global change: A call for policy action based on evidence from Queensland (Australia), Colombia and Brazil', *Global Environmental Change*, vol 19, pp21–33

Meinke, H. and Stone, R. (2005) 'Seasonal and inter-annual climate forecasting: The new tool for increasing preparedness to climate variability and change in agricultural planning and operations', *Climatic Change*, vol 70, no 1, pp221–253

Micheletti, M. (2003) *Political Virtue and Shopping: Individuals, Consumerism, and Collective Action*, Palgrave MacMillan, New York

Minx, J., Peters, G., Wiedmann, T. and Barrett, J. (2008) 'GHG emissions in the global supply chain of food products', paper presented at the International Input Output Meeting on Managing the Environment, Seville, 9–11 July

Müller, C. (2009) *Climate Change Impact on Sub-Saharan Africa? An Overview and Analysis of Scenarios and Models*, Deutsches Institut für Entwicklungspolitik, Bonn

OECD (Organisation for Economic Co-operation and Development) (2000) *Environmental Indicators for Agriculture*, OECD, Paris

Oosterveer, P. and Spaargaren, G. (2011) 'Organising consumer involvement in the greening of global food flows: The role of environmental NGOs in the case of marine fish', *Environmental Politics*, vol 20, no 1, pp97–114

Pirog, R., Pelt, T., Enshayan, K. and Cook, E. (2001) *Food, Fuel, and Freeways: An Iowa Perspective on How Far Food Travels, Fuel Usage, and Greenhouse Gas Emissions*, Leopold Center for Sustainable Agriculture, Ames

Pretty, J., Ball, A., Lang, T. and Morison, J. (2005) 'Farm costs and food miles: An assessment of the full costs of the UK weekly food basket', *Food Policy*, vol 30, pp1–19

Röös, E., Sundberg, C. and Hansson, P. (2010) 'Uncertainties in the carbon footprint of food products: A case study on table potatoes', *The International Journal of Life Cycle Assessment*, vol 15, no 5, pp478–488

Saunders, C. and Barber, A. (2007) *Comparative Energy and Greenhouse Gas Emissions of New Zealand's and the UK's Dairy Industry*, Agribusiness and Economics Research Unit, Lincoln

Soil Association (2007) 'Air freight consultation feedback statement', Soil Association, Bristol

Spaargaren, G. and Oosterveer, P. (2010) 'Citizen-consumers as agents of change in globalizing modernity: The case of sustainable consumption', *Sustainability: Science, Practice & Policy*, vol 2, no 7, pp1887–1908

Toulmin, C. (2009) *Climate Change in Africa*, Zed Books, London

Tukker, A. and Jansen, B. (2006) 'Environmental impacts of products: A detailed review of studies', *Journal of Industrial Ecology*, vol 10, no 3, pp159–182

UNCTAD (United Nations Conference on Trade and Development) (2009) *Trade and Development Report, 2009*, UNCTAD, Geneva

Van Hauwermeiren, A., Coene, H., Engelen, G. and Mathijs, E. (2007) 'Energy lifecycle inputs in food systems: A comparison of local versus mainstream cases', *Journal of Environmental Policy & Planning*, vol 9, no 1, pp31–51

Weber, C. and Matthews, H. (2008) 'Food-miles and the relative climate impacts of food choices in the US', *Environmental Science & Technology*, vol 42, no 10, pp3508–3513

# 6

# Local Food Provision

This chapter aims to:

- describe the widespread development of alternative, local agri-food networks;
- explore arguments for relocalizing food provision;
- discuss the dangers of romanticizing local food and oversimplifying requirements for sustainability.

## Introduction

Even as international trade in food products has increased rapidly, with greater roles for transnational agro-industrial corporations, 'big box' grocery stores and fast-food chains, so too have new local agri-food networks,[1] in a variety of forms and to an extent that lead many to invoke the formation of an alternative agri-food movement.

In these new, alternative agri-food networks, different social actors concerned about contemporary industrialized food provision find each other. *Small- and medium-scale farmers*, struggling to survive competition from large corporate farms, emphasize quality, freshness, local economic benefits and more. *Urban consumers*, interested in fresh, healthy foods, are flocking to farmers' markets, subscribing to box schemes with local farms for regular home delivery (CSAs), and planting home and community gardens. *Local food retailers*, including greengrocers and restaurants, are strengthening relationships with local growers. *Local governments and business associations*, seeking to boost both urban and exurban economies, have got into the act, offering facilities, staff and other support to launch farmers' markets and promote local agricultural 'brands' (appellations or *provenance*). *Food activists*, relating problems of poverty, malnutrition, obesity, diabetes and a host of other ills to poor access to healthy food, have launched urban gardening, 'farm to fork' and 'farm to school' programmes. All of these actors and many others have worked to support local agricultural production and consumption; create alternative, short food-supply chains; and provide a counterpoint to globalized food provision.

Participants in this alternative agri-food movement aim to address growing consumer concerns about food quality, human health, the environment, social justice and ethical dimensions of industrialized modes of food production and consumption. They argue that the awareness and social bonds necessary to strengthen social, economic and environmental sustainability can be recreated through directly connecting food producers, consumers, retailers, schools and other institutions.

In contrast to globalized food provision, in the case of local agri-food networks, sustainability is understood to include short supply chains, more fresh and seasonal food, and knowledgeable relationships between growers and consumers. Short supply chains demand less energy for transport, processing and packaging, while maximizing freshness and quality. When agricultural production practices are developed in concert with local ecologies and tastes, they arguably also optimize environmental impacts. Short food-supply chains buffer local producers, consumers and economies against the cyclicality of global markets, characterized by resource scarcities, oversupply problems, sectorial crashes and energy intensiveness. Local food-supply chains are diverse, may acquire various organizational forms and have few formal standards and procedures.

Though increasingly popular, alternative agri-food networks have been critiqued for romanticizing the local; not necessarily producing fresher, higher-quality food, with lower environmental impacts; and failing to address, or even exacerbating, local and global inequality. Some observers also comment that these alternative agri-food networks tend to ignore the necessity for continued national and international action to strengthen food sustainability and, overall, do not pose a viable alternative to global agri-food provision.

In this chapter, we first present the multifaceted phenomenon of these emerging alternative agri-food networks and then review their main characteristics. Criticisms are discussed in the final section.

## Local Food Resilience

A rapidly growing movement can be observed of small and medium-sized farmers, consumers, restaurant owners, local food retailers and others, creating new, local agri-food networks in counterpoint to the globalization of food provisioning (O'Hara and Stagl, 2001; Halweil, 2002; Green et al, 2003; Hines, 2003). Related initiatives include the development and marketing of local and regional food brands; the expansion and promotion of farmers' markets; direct sales to local restaurants, schools, hospitals and other institutions; new, urban agriculture; and the Slow Food movement. These different initiatives intend to create alternatives to the increased rationalization, industrialization and commercialization of food from 'farm to fork'. Alternative agri-food networks

### Box 6.1 Manifesto on the Future of Food

The 'Manifesto on the Future of Food' calls for 'a transition to a more decentralized, democratic and cooperative, non-corporate, small-scale organic farming as practiced by traditional farming communities, agroecologists, and indigenous peoples for millennia' (ICFFA, 2003, p4). It was formulated by the International Commission on the Future of Food and Agriculture (ICFFA), which is comprised of scientists and food activists, mostly from developed countries, but also from India and elsewhere.

*Source:* www.farmingsolutions.org/pdfdb/manifestoinglese.pdf (accessed 18 March 2011)

'tend to be place-based, drawing on the unique attributes of a particular bioregion and its population to define and support themselves' (Feenstra, 2002, p100) (see Box 6.1). Eating local is considered attractive because, among other reasons, the food is fresher and more flavourful; local growers are supported, strengthening local economies; and local food supplies are less vulnerable because they are protected from widespread food contamination, transportation problems and petrochemical price spikes and shortages.

## Local brands, varieties and regions

The second half of the 20th century saw the expansion, standardization and globalization of food and agriculture. These developments are still continuing. Pressed by low prices and high costs, small- and medium-sized producers around the world are letting marginally sustainable farms go to seed, turning where they can to other, often urban, sources of livelihood. New, suburban housing 'sprouts up' where crops have been grown for generations. Agricultural and varietal diversity are diminishing, while knowledge of traditional agro-ecosystems and the comfort and security of knowing where food comes from are disappearing for many consumers.

Against this still-rising urban tide, the early decades of the 21st century have witnessed a rediscovery and expansion of high-quality, healthy and unique local foods and food products. Traditional agricultural families and new pioneers have combined to develop high-quality local agricultural produce, products and regions – with support from local governments eager to identify new sources of revenue. Building on long-standing European traditions of wine (e.g. Champagne, Bordeaux), cheese (e.g. Camembert, Stilton) and other local food appellations, new food and drink regions have been trademarked and promoted by alliances of local producers, investors, governments, business organizations, co-operative extension agencies and others.

These efforts have emphasized developing new regions for high-value agricultural products (e.g. wine, cheese, maple syrup); high-quality, healthy food (freshly picked, organic); hospitality and 'agro-tourism' (wine-tasting, cheese tours, farm stays, crop mazes); and brand recognition and promotion, even in more distant markets. Proximity to major metropolitan areas can evidently be a plus for the success of these efforts. The Napa Valley wine-growing region, a short drive from the San Francisco Bay area in California, is one classic example. That region's success has been widely emulated in the US, from the Columbia River wine-growing region in Washington and Oregon, to the Finger Lakes wine (and more recently, cheese) region in New York, among many others.

With a focus on quality, taste, uniqueness and reduced environmental impacts, heirloom varieties have become increasingly popular. Against the few, standardized varieties of produce and other agricultural products offered in large quantities in 'big box' food retailers are tender, juicy heirloom tomatoes. Forgotten varieties and tastes of apples and dry-farmed, intensely flavourful produce are again available. And, agricultural products that are well suited to thrive (or at least survive) in particular, local agro-ecosystems are produced. Recast as extraordinary, high-quality products and experiences in a standardized world, local agri-food brands, varieties and regions are finding new legs to stand on.

## Coming soon, to a neighbourhood near you

In recent years, urban residents in North America and elsewhere have had to travel longer and longer distances to suburban 'big box' food retailers, increasing 'food miles' not only for food items, but also for themselves. As major food retailers abandon city centres, they leave behind 'food deserts': a dearth of fresh, healthy food in poor, urban neighbourhoods. The few, remaining small convenience stores charge high prices for not-so-fresh (if any) produce, reinforcing vicious cycles of poverty, unhealthy eating, obesity, diabetes and other food-related ills.

In these urban 'food deserts', new hope is inspired through the creation of neighbourhood and central farmers' markets. These vital, popular enterprises are capturing the imagination of local governmental and business officials, the media, producers and consumers in Europe, North America, Japan, Australasia and elsewhere. In the US in the last part of the 20th century, the number of farmers' markets grew more than ten-fold, from nearly 300 in the mid-1970s to more than 3000 by the end of the millennium. These alternative food-provision networks are based on two guiding principles: that the produce for sale is of local origin (in the UK, for example, with a maximum distance ranging between 30 and 75km) and that those selling the food should have been involved in its production.

Farmers' markets are one vehicle for reconnecting local agri-food producers – most of whom are too small and insufficiently connected to participate in large agri-food supply networks – with urban consumers. Working together, neighbourhood associations, schools, churches and others, local governments and food alliances have endeavoured to promote not only large central farmers' markets, but also smaller, neighbourhood farmers' markets, often on different days of the week in various locations. Local livelihoods and urban economies are boosted further through sales of produce grown in urban and community gardens.

## Box 6.2 The 100-Mile Diet

One popular manifestation of the relocalization of food provision is the '100-Mile Diet'. Organized by local-food advocates, churches and others, promoters of the '100-Mile Diet' call for consumers to purchase their food and agricultural products from sources located within 100 miles (161km) of their point of consumption. In such a diet, gone for most 'Northern' consumers are contemporary 'staples', such as olive oil, coffee, tea, bananas and chocolate, which are transported long distances across the globe. In their place are locally grown produce, dairy and meat products. As local agriculture in temperate zones is seasonal, the 100-Mile Diet is based also on relearning practices of canning, drying, freezing and other forms of food preservation. The 100-Mile Diet combines promotion of the local food supply with concerns about reducing the 'carbon footprint' of long food-supply chains, through shortening distances that food and food products have to travel between 'farm and fork', thus reducing 'food miles'. Often part of a regional development strategy, the 100-Mile Diet also relates to the Slow Food Movement.

*Sources:* www.100milediet.org; Smith and MacKinnon (2008) and Kingsolver et al (2008)

Shoppers visit farmers' markets for a variety of reasons. Food quality and price are important. But so, too, are the local and social *embeddedness* of exchange, including direct, face-to-face relationships between consumers and producers (Hinrichs, 2000; Halweil, 2002, 2006; Kirwan, 2004), a close connection with regional origin or *provenance* and sometimes with traditional local cultures. The symbolic meaning of this food is imbued in such values as 'authenticity' and independence from globalized, 'industrialised, chemical-dependent ... mass food production systems' (Seyfang, 2007, p109) (see Box 6.2). Many of these elements tend to be missing from globalized food provision and 'big box' food retailing.

### Farms in the city?

Although centuries old, urbanization is an important social process as ever, with, for the first time in the 21st century, a majority of the world's population living in cities. How to feed these rapidly increasing numbers of urban residents is a critical problem around the globe. Each city's growth is particularistic, dependent on various specific geographic, demographic, economic and other factors. In North America, older cities in de-industrialized regions face problems of loss of livelihoods, poverty, decaying infrastructure and population loss. In newly industrializing regions, including in Asia, Africa and Latin America, cities face rapid in-migration, informal housing, economic insecurity and more. In both older and newer industrialized countries, urban agriculture is becoming increasingly important for food security, livelihood and health.

A fascinating film[2] on urban agriculture in the developing world cites a UN Development Programme (UNDP) estimate (Smit et al, 1996) that, 'worldwide 800 million urban residents are engaged in agriculture'. Agriculture in cities takes place in backyards, in school lots, on roof tops and alongside public infrastructure projects, such as roads and waterways. Often, urban edges serve as intensive 'truck' gardens, growing fresh produce for city dwellers. Where employment is

## Box 6.3 Organic vegetables from the slums

In informal settlements around Capetown, South Africa, one can find thousands of organic vegetable gardens, both private and community-organized. Supported by Abalimi Bezekhaya ('Farmers of Home'), a local urban agriculture and environmental action association, residents produce their food organically because it is 'easier and less expensive', and since they cannot read, it would be difficult for them to safely follow printed directions for pesticide applications anyway. With fresh, local produce, residents have ready access to inexpensive vegetables; and by selling surplus through organic shops elsewhere in the metropolitan area, they can earn additional cash.

*Source:* Petit-Perrot (2009); see also www.abalimi.org.za (accessed 18 March 2011)

scarce and low-waged, especially for newly arrived urban residents, backyard and other forms of urban farming provide important economic as well as nutritional supplementation. In another recent study (Zezza and Tasciotti, 2010, summarized in FAO, 2010, p1), the authors find that, 'based on data from 15 developing and transition countries ... up to 70 per cent of urban households participate in agricultural activities'. They suggest that 'urban agriculture seems particularly important in low income countries such as Malawi, Nepal and Vietnam' (see Box 6.3, for an example from South Africa).

Urban agriculture is also expanding in more developed countries.[3] Former manufacturing hubs, such as Detroit, are losing jobs and population while they are beset by suburbanization as well. Older, inner-city residential neighbourhoods fall into disrepair, with dilapidated homes boarded up, subject to drug users, squatters and arson, and ultimately bulldozed away. More and more vacant lots appear in the city, near perennially impoverished residents. For local anti-poverty and food activists, however, this has become an opportunity to rediscover urban agriculture. In a series of reports in the *Guardian* (UK) newspaper, Paul Harris has documented Detroit's urban agricultural renaissance. According to Harris (2010), the city's population fell by half, 'from about 1.8 million at its peak in the 1950s to fewer than 900,000 today'. During his visits, he found 'thousands of people involved in urban farming in Detroit'. These new urban farmers have plenty of land: 'Abandoned houses, vacant lots and empty factories now make up about a third of Detroit, totalling around 40 square miles – the size of San Francisco'. He finds both community groups and commercial interests actively exploring urban agriculture. These initiatives have 'certainly caught the attention of cash-strapped local government'.

Urban agriculture contributes to people's health and well-being and minimizes the ecological impacts of food provision by eliminating long-distance transport from rural areas and by reusing organic waste through composting.[4] Urban agriculture can be used as a building block for a more encompassing urban food strategy as well. In recent years, a growing number of cities have been redesigning food provision with a wider perspective (Sonnino, 2009). An important complement to urban agriculture is how public and private institutions organize food procurement. Schools, child care centres, hospitals, nursing homes, prisons and other large institutions all regularly procure large volumes of food for their charges. In all of these settings, stakeholders (parents, students, patients, inmates, families, health officials and others) may lobby for local food provisioning.[5]

## 'Farm to fork' networks

Among the innovators and advocates for relocalizing food provision are local restaurants featuring fresh, seasonal, locally grown, often-organic produce. Famous restaurateurs and chefs, including Alice Waters at *Chez Panisse* in Berkeley, California, have developed direct relationships with local, organic growers, who provide fresh, wholesome produce daily. Others, such as chef-owner Alicyn Hart at *Circa* in Cazenovia, New York, bring farmers' markets into their establishments, offering customers not only hot dishes but also local 'cheeses, cage-free eggs, organic dairy products' and more. Led by Waters and others, local-food advocates have been working closely with primary and middle-schools, developing vegetable and herb school gardens.[6] These activists are re-evaluating school-lunch programmes that over the years have become increasingly centralized and 'industrialized' along fast-food models, with lots of high-fat, processed foods: fried 'chicken tenders', frozen pizzas, mass-produced beef patties, etc. In addition to school gardens, advocates have initiated 'farm to school' programmes, linking schools with local farms. In this way, not only do students learn more about where their food comes from, but the food is fresh, high quality and locally produced, helping sustain local farmers and strengthen local economies.

Along similar lines, in their publications, Morgan and Sonnino (Morgan and Sonnino, 2008; Sonnino, 2010) stress the benefits of promoting local provisioning of school meals. In many places, governmental authorities (at local, regional or national levels) are legally obligated to provide nutritious meals for school children. Many initially regarded this as a strategy to improve school children's health and educational performance, but over time it became just another public service to be privatized against lowest costs. As a result, in the 1990s and 2000s, many school meals were of low quality, based on industrialized and cheap food, contributing to problems of obesity among today's youth. In response, in different regions, local authorities or parent associations undertook actions to improve school meals by developing procurement policies that promoted local and healthy foods. In many instances, these groups were able to substantially improve the quality of the school meals supplied, in combination with strengthening networks between schools and local farmers and food processors. The promotion of locally supplied school meals has also exposed children to a greater diversity of fresh food products.

## Subscription farming

For a time, small- and medium-scale farmers were able to carve out economically viable niche markets in high value-added organic produce. After decades

## Box 6.4 Growing Power in Milwaukee

On the northern outskirts of Milwaukee, Wisconsin stand 14 greenhouses on almost 1 hectare of land. 'Growing Power' is the name of a farm producing many different food products, a food-distribution hub in an area without full-service grocery stores, and a training centre. Included in the complex are an aquaponics fish-raising facility, an apiary, three poultry houses for laying hens and ducks, an anaerobic digester to produce energy from the farm's food waste and more. Every week, Growing Power delivers up to 350 baskets of food for distribution in the local community. Together with hundreds of customers, the staff of 35 become familiar with growing and eating fresh produce, an opportunity they would not get otherwise.

*Source:* Bybee (2009); www.growingpower.org

of extraordinary growth in the natural, healthy and organic foods sector, agro-industry and major food retailers alike saw gold, making major investments and converting large-scale farms from conventional to organic agriculture. The niche carved out by small-scale farmers became increasingly competitive, however, as it was threatened by the entry of agro-business. Economies of scale continued to support large-scale rather than smaller, local producers. In response, by the end of the 1990s, local farmers developed new initiatives, emphasizing the importance not only of organic agriculture, but also of sustaining local farms, economies and food security. Among their new set of survival strategies are CSAs, which are subscription or membership schemes, in which mostly urban consumers make subscription purchases in advance of one or more of a farm's outputs. CSAs allow local growers to generate financial capital to run their farms, and in return, when the harvests come, shareholders receive their shares of the produce. Schemes range from 'rent a cow', shares of milk produced by a female bovine, to 'bird of the month', the advance purchase of roasting fowl (priced to include feed, water, shelter and a margin of profit for the farmer) and 'bouquet of the week', fresh cut flowers delivered each week. There are also schemes that guarantee a weekly delivery of free-range eggs, and many other such agreements (see Box 6.4). The US Department of Agriculture identified more than 12,500 farms engaged with more than 270,000 households through CSAs in 2007 (USDA, 2009, p606). By sharing the costs and harvests of agriculture, farming risks and rewards are more equitably balanced.

### Alternative labour strategies

The quality, affordability and availability of farm labour remains one of the biggest challenges to small farmers today. There are more and more instances where crops have been planted or fruit trees are ready to harvest, but labour costs are so high that produce is left unharvested in the fields or fruit remains in the trees. Growers have addressed this in a variety of ways. Some in high-wage OECD countries have resorted to using immigrant, even undocumented, farm labour. Sometimes even such sources of labour are unavailable, however. One alternate labour strategy employed by many small-scale farmers is self-pick (or 'u-pick') schemes, whereby growers plant fruits and vegetables and customers do the harvesting. Consumers thus gain access to food at prices discounted over what they might find – imported from far-away places – in local grocery stores. In another approach in North America and elsewhere, where retired and unemployed people have an 'excess' of labour and a 'deficit' of nutrition, urban-food advocates have organized 21st century 'gleaners brigades' to harvest and distribute produce that otherwise would rot in the fields. One such, religiously oriented group, the Society of St Andrew, based in Virginia but with efforts across the US, claims to have involved 'more than 30-thousand volunteers ... to salvage and distribute ... more than 18-million pounds of produce through our Gleaning Network'.[7]

### Slow Food

Starting in Italy, a country known worldwide for its small enterprises and long-standing regional agricultural specialties, a new Slow Food movement has risen. Inspired by its charismatic leader, Carlo Petrini, the Slow Food movement spread rapidly, especially in more affluent parts of the world, as an alternative to fast food, symbolized by golden arches. Slow Food combines food provisioning, food preparation and food consumption as practices that are simultaneously social and biophysical in nature, with an emphasis on unique quality, taste, sociality, identity and tradition, rather than on quantity, quickness and uniformity. The movement has spurred many local initiatives worldwide, but has become especially active in Europe, the US and Japan. The markets for many, sometimes even forgotten, local and regional food products have been strengthened (Morgan et al, 2006). Small, local agricultural producers are elevated as champions of tradition, quality and community. Communities become deeply embedded in slow food products through reputation, provenance, appellation, marketing and practice. The universal, industrialized fast-food model is rejected in favour of particularistic, small-scale, locally and community-based food production and consumption that centres on the human relationships of farmers with their farms and consumers with their food, and also those between producers and consumers. Therefore, the Slow Food movement can be considered primarily a cultural critique of global, fast food.

## The Argument for Local Food

As noted above, a loose but widespread alternative agri-food movement is challenging the increased globalization of food and agriculture. Having identified some of the main manifestations of this movement, we now review some of its major claims. Among these arguments, we find that its promoters consider short food-supply chains to be more ecologically sustainable than long ones; that seasonal, locally produced foods are fresher and healthier; and that local food provisioning strengthens social bonds and communities. Other arguments hold that the viability and sustainability of local economies are enhanced through reduced dependence on global agri-food systems, and that the diversity of plants, animals, producers, products, consumers and tastes is conserved and even enhanced through sustaining local agriculture.

## Short supply chains

Globalized food provisioning requires transporting agricultural goods over very long distances, from one part of the globe to many others. Fresh foods grown elsewhere in the world demand refrigeration, climate control and dependable, quick transportation to maintain freshness and quality. Many agricultural products are processed (frozen, freeze-dried, canned, packaged, etc.) prior to shipping. In addition to affecting nutrition, these processes are also energy-intensive and often polluting to local water supplies. Very large-scale, often vertically integrated, globally sourced corporate farms and agro-processing facilities may in some respects be highly efficient, but overall, they arguably require higher energy inputs because of greater transportation, refrigeration, climate control and processing costs. This model of food provision also results in the unwanted, locally intensive disposal of agricultural and food-processing wastes. By contrast, a greater proportion of local provisioning means food that is fresh is brought to local markets more quickly and has a higher nutritional quality with less associated processing energy costs and waste. Further, when local agricultural-production practices match the ecological characteristics of a particular region, especially its seasonal rhythms, they will have optimal overall environmental impacts.

## Fresh and seasonal food

The freshness, taste and nutritional value of many foods are greatest at the moment of their harvest or primary processing stage. The longer the supply chain, the more processing, preservatives and refrigeration are required to maintain the food's looks of freshness, not only adding to food costs, but also reducing overall food quality, taste and nutrition. Local provisioning moves food most quickly and nutritiously from 'fields' to 'fork'. Restaurant owners know this well, and thus they contract local growers, wholesalers or greengrocers to deliver fresh produce daily, or even better, grow their own herbs and other foods in kitchen gardens. Food tastes best and is healthiest when fresh. For many food items, fresh and raw items have the highest nutritional content. The 'natural' character of the products and production methods used in local food-supply networks is considered the best guarantee for protecting the health of human beings, farm animals and the ecosystem as a *whole* (Green et al, 2003). Still, the foods that fit such methods of supply represent just one part of the ordinary diet of most present-day consumers, and to profit optimally from the qualities of local food, this diet would also have to change.

## Building relationships

Local food-supply chains are very diverse and can acquire very different organizational forms because there are no overarching formal standards and procedures. Locally organized, often community-based food-supply chains 'tend to be place-based, drawing on the unique attributes of a particular bioregion and its population to define and support themselves' (Feenstra, 2002, p100). Local communities or networks may develop their own procedures and requirements, making this a very flexible arrangement. Nevertheless, although the organizational forms developed within various local food-supply networks may differ considerably, active consumer participation is an essential characteristic of each of them.

Localizing food provision can also be considered a form of social action, an alternative to mainstream food provision. According to such a view, mainstreaming local food would be seen not as a positive goal, but rather as co-optation or 'conventionalization'.[8] According to this perspective, local food has a symbolic meaning that challenges the main, globalized industrialized food system, because the former's values and structures indicate that other options are possible. The presence of alternative, local food-supply chains provides tangible evidence that globalization is not the only option. By participating in locally oriented forms of food consumption, people may join a social movement (Melucci, 1996) that addresses relevant social problems and expresses alternative values and aspirations. Producer and consumer participation in different aspects of local food provision creates space for creativity, where alternative social and economic practices can be introduced. Local food is selected not only because of its inherent characteristics, but also because of its symbolic meaning, which consumers identify as alternative and something that they can use politically (Terragni et al, 2009). Thus a distinction can be made between local food as a *political agenda*, directed at creating an alternative food economy, and as a *development strategy* to incorporate small rural firms and marginal agricultural economies into economic development (Fonte, 2008). Promoting the first goal means reconnecting producers and consumers, and the second goal means primarily creating regional food identities to strengthen their position on the larger market for food.

Active consumer participation is an essential characteristic in many of these modern, localized supply chains. The involvement of consumers makes it possible to establish trust in the quality and safety of food based on personalized, face-to-face contacts, without relying on expert-based systems, as is necessary in abstract global food-supply chains. Becoming involved in local food-supply chains can promote a sense of community awareness and integration, as networks between like-minded people are strengthened (Watts et al, 2005).

## Food security

The small scale of local food-supply chains allows small farmers, local communities and consumers some independence from globalized agri-food systems, which are deemed unsustainable over the long haul. Only alternative local food-supply chains are considered capable of responding to contemporary

## Box 6.5 Community food security

In their *Guide to Eating Locally and Seasonally*, the Interfaith Network of Portland, Oregon, suggests that:

> *Food has the potential to weave together the land, people and communities into a fabric of relationships that fosters justice and sustainability. Unfortunately, the food we eat often represents unjust relationships and a degraded Earth. Community food security (defined as all persons in a community having access to fresh, local, culturally appropriate food at all times) is a concept and process that creates healthy relationships around food.*

*Source:* Interfaith Network (2003)

consumer concerns about health, the environment, animal welfare and social impacts of modern practices in food production and consumption. According to Princen (1997), only de-globalization through the creation of small-scale local food networks would enable the checks and balances necessary for *sustainable resource use of local land and other agricultural resources*, as local networks function on the basis of direct interaction between food producers and consumers. By developing alternative networks, local producers guard themselves against being subordinated in spatially extensive food-supply chains and reduce their vulnerability to subordination within these chains, which serve the interests of powerful actors (Watts et al, 2005). Local food is seen as countervailing the ever-expanding market shares and increasing profits that large corporations are continuously making efforts to expropriate (see Box 6.5).

In response, local food-supply systems offer market opportunities for small (organic) producers and, together with consumers, co-produce feelings of enchantment (Thompson and Coskuner-Balli, 2007). These initiatives differentiate themselves by re-territorializing the food market and blurring the traditionally strict difference between producers and consumers as market actors by making them co-producers of the food system. In some respects, the concept of time is also changing because pre-modern images of farming and rural life are merged with the newest technologies, especially in marketing.

'De-globalizing' food production and consumption and creating localized systems of food provisioning may offer attractive opportunities for innovative forms of governance as well. The nation state has no need to interfere with interactions at the community level, as producers and consumers take up these responsibilities themselves. Involvement by the nation state is needed only to protect these initiatives from domination by large corporations and its harmful effects and, in general, to support the principles of a localized agro-ecology to ensure local food security in combination with the vitality of robust rural economies. At the same time, this focus on optimizing local resource use may reproduce inequalities in the availability of such resources between different localities. Promoting local food security may therefore put pressure on food security at the global level.

## Supporting local diversity

Scholars studying the modernization of agriculture in Europe and elsewhere have documented the persistence of a high level of diversity in farming practices, despite the standardization and homogenization expected from globalization and industrialization (Van der Ploeg et al, 2000; Renting et al, 2003). Diversity is not just a relic of the past; it can be seen today in highly specialized local farms as well. Globalization notwithstanding, there remains room for multi-level, multi-actor and multifaceted ways of providing food. Van der Ploeg et al (2000, p399) consider the emergence of new localized food-supply chains as 'maybe the most evident example of the reconfiguration of resources and networks in rural development'. Diversity in contemporary farming practices means not only that local food systems are still heterogeneous, but also that producers and consumers may negotiate understandings of those systems' material and symbolic meanings in different ways. As Selfa and Qazi (2005) argue, in urban areas, producers and consumers identify local food with physical proximity, while this is not necessarily the case in more rural areas. Localizing food provision can evolve along different lines and create various configurations depending on time and place. Local agri-food networks can be a starting point for sustainable development as it may contribute to creating 'more resilient and robust regions' (Wiskerke, 2009, p383).

# Critical Perspectives

In a rejoinder to the idealistic and sometimes romantic promotion of (re-)localized food provisioning, critics advance several provocative questions, including: are localized, short food-supply chains sufficient to feed the rapidly growing urban populations around the world? In this era of hyper-globalization, are there any truly 'local' agri-food products? (And if so, are they not threatened by new global challenges, such as climate change, invasive agricultural pests and diseases and more?) By strengthening local food-supply chains in the global North, are not advocates and consumers unfairly closing attractive export markets to agricultural producers in the global South?

## A real alternative?

Critics commenting on the promotion of local food supply claim that being physically close does not necessarily result in less market orientation and instrumentalism in the relationships between producers and consumers (Hinrichs, 2000). Consumers may not act differently when buying food at a farmers' market compared to buying it in a supermarket. Furthermore, sustainable food production can be organized in different ways, not only through the physical distance between food producers and consumers (Scoones and Toulmin, 1999; Evans et al, 2002; Keeley and Scoones, 2003). Also, feedback between food producers and consumers regarding the sustainability impact of practices and behaviours may be considered to be of a social and not a mechanical character. Feedback must be organized by thematizing the sustainability impact and creating social pressure for implementing solutions that are identified, and this is not automatically generated. At the same time, however, global food-supply chains may not be as disengaged from local social and ecological dynamics as proponents of local food provision often suggest. Producing food remains to be realized under natural conditions that can be manipulated only to a certain extent; if ignored too much, ecological and social dynamics may have unexpected impacts on industrialized food production systems (Morgan et al, 2006).

Despite the popularity of locally organized food-supply chains, such a process of de-globalization does not necessarily provide the only response to the challenges facing contemporary food production and consumption practices today (Evans et al, 2002). As a large part of the food consumed by consumers in OECD countries is already of global provenance, it is important to also seek options for increasing the sustainability of these globalized food-supply chains. If globalization is approached as a heterogeneous and complex process, rather than as a uniform and homogeneous one, other innovative governance arrangements of global food provision may also provide useful and effective responses. Through their focus on the ecological consequences of global supply chains, proponents of local food supply risk ignoring other dimensions of sustainability, such as the social and economic consequences for producers in developing countries. If these broader considerations of sustainability were to be integrated in international trade, the global food supply could be considered to have positive impacts.

In contrast, Seyfang (2007) argues that, at present, local food may not yet be a solution for contemporary food provision, not because of its inherent characteristics, but because mainstreaming successful niche experiments requires institutional support, which is currently lacking. Many consumers are interested in experimenting with alternative lifestyles and relying more on local food, but current agricultural and food policies and institutions limit the expansion of alternative systems. Legal requirements, unfair subsidy arrangements and a general lack of support make the future perspective of local food-supply systems difficult. To consolidate innovative practices, participants in local food systems need to feel that they are part of larger communities that share their values. Stronger political support could facilitate the growth of such a movement.

## Embeddedness

Local food is held to be different, embedded in local scale and better than larger-scale food provision. Local food is expected to be more sustainable, more socially just and democratic, delivering better nutrition, safety and quality. Local food systems promote 'socially embedded economies of place' (Seyfang, 2006, p386) that promote personalized trust between producers and consumers and strengthen local economic development. Lyson (2005) considers what he calls *civic agriculture* to be embedded in local communities because it contributes to community health and vitality. In relying on site-specific knowledge about

## Box 6.6 Local trap?

In his recent book, *Ecological Intelligence*, Daniel Goleman recounts:

> *Life Cycle Assessment raises the question, what exactly do we mean by 'local'? A Montreal-based industrial ecologist tracked the geography of the life cycles of tomatoes grown in greenhouses near Montreal. As she told me, 'Not much local was in the "local" product. While the tomato R&D was conducted in France, the seeds were grown in China and transported back to France, where they were treated and shipped to Ontario, where the seedbeds were sprouted. Finally, these seedbeds are trucked to Quebec, where the final plant is cultivated and the fruit harvested. Even a "local" tomato has a global past'.*

*Source:* Goleman (2009, p55–56).

farming, local food farmers can produce higher-quality food for consumers, who now have a voice in how and where their food is produced. Local food thus becomes a focal point around which non-market relationships between previously distanced persons, groups and institutions can be built.

Born and Purcell (2006) criticize this view as being *a local trap* (see Box 6.6). They argue that there is nothing inherent about any scale, because the impacts of a food system depend on the actors and agendas empowered by the particular social relations present within a certain food system. Local food systems may therefore have negative as well as positive outcomes for local food provision because these outcomes depend on the particular actors involved, as well as their strategies and practices. In the authors' view, localization should not be privileged as a goal for food strategies in general, but seen rather as a goal of those empowered by a particular scalar strategy. The scale of food provisioning, whether local, global or regional, is socially constructed, without having any inherent ethical (or sustainability) characteristics. From Born and Purcell's perspective, food strategies therefore should be analysed on the basis of who is empowered, rather than of the scale at which they are enacted.

Winter follows a similar line of argument when he states that 'all market relations are socially embedded' (2003, p25). Relations between producers and consumers require mutual trust but also hold meaning (the signification given to particular purchases). Whether local food heralds a more ecologically sound provision system is therefore an empirical question.

In response, some authors suggest that this argument is too general and that the advantages resulting from local food provision are not just founded on general principles but also in empirically confirmed cases. Research has shown that localism and devolution can effectively create 'deeper democratic structures, social and spatial solidarity and sustainable development' (Sonnino, 2009, p15).

### Defining 'sustainability'

In another critique, 'local' should not be conflated with 'sustainable', as the positive impacts of consuming local food are overshadowed by the negative

## Box 6.7 Exploring environmental consequences of local food production

Almere is a rapidly growing town in the Netherlands, and municipal authorities want to promote sustainability in one of its new neighbourhoods, including by increasing local food production. In a scenario study, experts determined the consequences of producing some 20 per cent of the total food basket. The chosen 20 per cent represents the proportion of current diets that could be produced locally, given weather conditions, etc. The calculated ecological effects of such a change are relatively small and primarily determined by the chosen technology in primary production (i.e. use of renewable energy or not). This choice is independent of the production location, however. A more important impact depends on the distribution system, because if it is well organized, with a large number of distribution centres, the total number of kilometres driven could be reduced by nearly 90 per cent. The study concluded that this shift is possible, but that larger effects would require a reduction of animal proteins in the diet and not just localizing food production.

*Source:* Sukkel et al (2010)

impacts. The designation 'local food' itself does not indicate what processing methods have been used, in contrast to organic food, for instance. Local food refers only to the physical distance between the locations of production and sale (Watts et al, 2005), but the environmental impact of food depends not only on the distance it has travelled; it is also influenced by the ways in which it has been produced, processed and transported. For instance, local food may use less energy for transport but more water and land, and in general, its production methods may use natural resources less efficiently than other methods. Local foods therefore should not necessarily be conflated with being organic and better tasting, and with saving family farms and strengthening local communities, because this all must be proven in particular instances.

Increasing the scale can improve the energy efficiency of production, processing and logistics activities, and effective use of available natural resources (sunlight, water, seasons, etc.) can reduce the ecological impact of producing food. It is therefore important to rely not only on a common-sense approach to sustainability but to take trade-offs into account in a comprehensive manner (see Box 6.7).

### Impacts on the poor

'De-globalizing' food provision in the developed world would not necessarily address agri-food-related problems of poverty and environmental degradation in developing countries. Demanding 'self-reliance' in food provision in the North would deny countries in the South the possibility of exporting agricultural products to rich country markets and only add to the already-existing poverty and inequality between the rich and the poor. Kevin Watkins (2002) from Oxfam suggests that:

> *if trade is to work for the poor, we need to challenge the power relations and vested interests that make markets work for the rich. That means putting land redistribution, workers' rights, environmental sustainability and the curtailment of corporate power at the heart of the agenda.*[9]

Safeguarding opportunities for the poor in developing countries to access markets in richer countries could then be considered a contribution to sustainability and equality instead of a diminution. Marsden (2004, p138), for example, claims that:

> *what marks alternative food chains out from the conventional system is by no means their face-to-face nature necessarily. In some of the more mature quality supply chains we see the development of spatially extended networks, which are selling brands and labels and seriously commodifying their culinary repertoires (e.g. Parmigiano Reggiano Cheese). They are still categorically alternative, however, in that they have done and do re-equate nature, space, socio-technical practices, and quality conventions in ways which make it impossible to replicate these outside that network. These then are the new ecologically deepened supply chains.*[10]

In developed countries, local food-supply approaches are criticized for leading to another unintended consequence for the poor: they may create divides between richer and poorer consumers. The 'local' may transform into just another market segment allowing for higher added value than conventional food. For instance, farmers' markets are mostly located at the more attractive sites from sellers' perspective, i.e. in more affluent neighbourhoods. Overall, local food is often more expensive than the food poor people normally buy, while they may also lack the cultural resources required for its preparation.

Proponents of local food have mostly pointed out the need to promote alternative agri-food networks at different locations to support the poor. Some of the examples presented in this chapter are used to support the possibility of this claim.

## Conclusion

Alternative local agri-food networks are rapidly growing in number and developing into recognized alternatives to conventional industrial global food-supply systems. In this chapter, we provided illustrations of this trend and discussed the contributions this alternative may offer for future sustainability in food provision. Local supply chains are attractive because they try to optimize available diversity, reduce energy needs and prevent problems of oversupply or scarcity through direct communication between producers and consumers, thereby restricting the role of market dynamics in the food supply. We found that local agri-food networks may have diverse impacts on sustainability and that uniform optimism in this respect is not justified. Next, we also observed that in such agri-food networks, local identities are continuously created and recreated, flexible and mouldable, rather than fixed. Local agri-food networks are also heterogeneous social movements, bringing together people concerned about the consequences of the contemporary way of industrialized food provision, including farmers and activists fighting corporate farming and protecting local communities, consumers who look for fresh and healthy foods of high quality with low environmental impact, and concerned citizens trying to reduce malnutrition (including obesity) among poor people without adequate access to healthy food. The identity of local agri-food networks is primarily built around their protest character, whereby the local is seen as a site of resistance to the global, even while the site itself is also created through globalization. It may therefore be more profitable to consider local agri-food networks through their interactions with other, including global, food-provision practices than in isolation. Through interacting with global food networks, the meaning of 'local' becomes blurred, as it is filled with all kinds of different notions and values. There is a serious danger that local agri-food networks may become overburdened with too many heterogeneous expectations that they cannot meet. It therefore may be more helpful to approach local agri-food networks as part of more encompassing food-supply networks and accept heterogeneity and diversity within them than aiming for their purification.

---

*Take-home lessons*

- Alternative, local agri-food networks provide new sources of hope, nutrition and livelihoods for producers and consumers.
- Urban agriculture, in a variety of novel forms, is a critical part of food provisioning for millions worldwide.
- Global and local food sourcing may both be necessary to feed the world and provide livelihoods.

---

## Notes

1. Also called 'agro-food' networks.
2. Video on policy implications and examples of urban agriculture by Resource Centres on Urban Agriculture and Food Security (RUAF), www.ruaf.org/ (accessed 18 March 2011). The film explores 'intra-urban' and peri-urban' agriculture in Ecuador, Tanzania, Senegal and Viet Nam.
3. See www.cityfarmer.info (accessed 18 March 2011) for many more experiences from different cities around the world.
4. See the work by the International Development Resource Centre (IDRC), which has a section, called 'Working with urban farmers for food security' on its website (www.idrc.ca/in_focus_cities, accessed 18 March 2011).

5 For a discussion of experiences in one of the most elaborate urban food strategies, Bela Horizonte in Brazil, see Rocha and Lessa (2009).
6 For instance, US First Lady Michelle Obama's engagement in kids growing their own food in gardens at school. See the US National Gardening Association (www.kidsgardening.org/white-house-garden, accessed 18 March 2011).
7 See www.endhunger.org (accessed 18 March 2011).
8 See Chapter 9 on the conventionalization thesis in US organic agriculture.
9 Retrieved from www.maketradefair.co.uk/en/index.php?file=28052002092914.htm (accessed 16 March 2011).
10 See also Goodman (2004).

## Further Reading

Morgan, K., Marsden, T. and Murdoch, J. (2006) *Worlds of Food: Place, Power, and Provenance in the Food Chain*, Oxford University Press, Oxford: offers an insightful overview of the relevance of geographical scale in contemporary food provision; contains several case studies, including some on local food.

## References

Born, B. and Purcell, M. (2006) 'Avoiding the local trap: Scale and food systems in planning research', *Journal of Planning Education and Research*, vol 26, pp195–207

Bybee, R. (2009) 'Growing power in an urban food desert', *Yes! Magazine*, 13 February

Evans, N., Morris, C. and Winter, M. (2002) 'Conceptualizing agriculture: A critique of post-productivism as the new orthodoxy', *Progress in Human Geography*, vol 26, no 3, pp313–332

FAO (Food and Agriculture Organization of the United Nations) (2010) *Fighting Poverty and Hunger: What Role for Urban Agriculture?*, Department of Economic and Social Development, FAO, Rome

Feenstra, G. (2002) 'Creating space for sustainable food systems: Lessons from the field', *Agriculture and Human Values*, vol 19, pp99–106

Fonte, M. (2008) 'Knowledge, food and place: A way of producing, a way of knowing', *Sociologia Ruralis*, vol 48, no 3, pp200–222

Goleman, D. (2009) *Ecological Intelligence: Knowing the Hidden Impacts of What We Buy*, Allen Lane, London

Goodman, D. (2004) 'Rural Europe redux? Reflections on alternative agro-food networks and paradigm change', *Sociologia Ruralis*, vol 44, no 1, pp3–16

Green, K., Harvey, M. and Mcmeekin, A. (2003) 'Transformations in food consumption and production systems', *Journal of Environmental Policy & Planning*, vol 5, no 2, pp145–163

Halweil, B. (2002) *Home Grown: The Case for Local Food in a Global market*, Worldwatch Institute, Washington, DC

Halweil, B. (2006) *Good Stuff? Local Food*, www.worldwatch.org/node/4132, accessed 28 June 2006

Harris, P. (2010) 'Detroit gets growing', *Guardian*, www.guardian.co.uk/environment/2010/jul/11/detroit-urban-renewal-city-farms-paul-harris, accessed 18 March 2011

Hines, C. (2003) 'Time to replace globalization with localization', *Global Environmental Politics*, vol 3, no 3, pp1–7
Hinrichs, C. (2000) 'Embeddedness and local food systems: Notes on two types of direct agricultural market', *Journal of Rural Studies*, vol 16, pp195–303
ICFFA (International Commission on the Future of Food and Agriculture) (2003) *Manifesto on the Future of Food*, ICFFA, San Rossore
Interfaith Network (2003) *Portland's Bounty: A Guide to Eating Locally and Seasonally in the Greater Portland and Vancouver Areas*, www.emoregon.org/inec_food.htm, accessed 15 December 2003
Keeley, J. and Scoones, I. (2003) *Understanding Environmental Policy Processes: Cases from Africa*, Earthscan, London
Kingsolver, B., Hopp, S. and Kingsolver, C. (2008) *Animal, Vegetable, Mineral: A Year of Food Life*, Harper, New York
Kirwan, J. (2004) 'Alternative strategies in the UK. Agro-food system: Interrogating the alternity of farmers' markets', *Sociologia Ruralis*, vol 44, no 4, pp395–415
Lyson, T. (2005) 'Civic agriculture and community problem solving', *Culture and Agriculture*, vol 27, no 2, pp92–98
Marsden, T. (2004) 'The quest for ecological modernisation: Re-spacing rural development and agri-food studies', *Sociologia Ruralis*, vol 44, no 2, pp129–146
Melucci, A. (1996) *Challenging Codes: Collective Action in the Information Age*, Cambridge University Press, Cambridge
Morgan, K., Marsden, T. and Murdoch, J. (2006) *Worlds of Food: Place, Power, and Provenance in the Food Chain*, Oxford University Press, Oxford
Morgan, K. and Sonnino, R. (2008) *The School Food Revolution: Public Food and the Challenge of Sustainable Development*, Earthscan, London
O'Hara, S. and Stagl, S. (2001) 'Global food markets and their local alternatives: A socio-ecological economic perspective', *Population and Environment*, vol 22, no 6, pp533–554
Petit-Perrot, C. (2009) 'Organic vegetables from the slums', *Afrika Nieuws*, vol 15, October
Princen, T. (1997) 'The shading and distancing of commerce: When internationalization is not enough', *Ecological Economics*, vol 20, pp235–253
Renting, H., Marsden, T. and Banks, J. (2003) 'Understanding alternative food networks: Exploring the role of short food supply chains in rural development', *Environment and Planning A*, vol 35, no 3, pp393–411
Rocha, C. and Lessa, I. (2009) 'Urban governance for food security: The alternative food system in Belo Horizonte, Brazil', *International Planning Studies*, vol 14, no 4, pp389–400
Scoones, I. and Toulmin, C. (1999) *Policies for Soil Fertility Management in Africa*, IIED, IDS and DFID, Edinburgh, Brighton and London
Selfa, T. and Qazi, J. (2005) 'Place, taste, or face-to-face? Understanding producer–consumer networks in "local" food systems in Washington State', *Agriculture and Human Values*, vol 22, no 4, pp451–464
Seyfang, G. (2006) 'Ecological citizenship and sustainable consumption: Examining local organic food networks', *Journal of Rural Studies*, vol 22, pp383–395
Seyfang, G. (2007) 'Cultivating carrots and community: Local organic food and sustainable consumption', *Environmental Values*, vol 16, pp105–123
Smit, J., Ratta, A. and Nasr, J. (1996) *Urban Agriculture: Food, Jobs and Sustainable Cities*, UNDP, Geneva and New York

Smith, A. and MacKinnon, J. (2008) *Plenty: Eating Locally on the 100-Mile Diet*, Three Rivers Press, New York

Sonnino, R. (2009) 'Feeding the city: Towards a new research and planning agenda', *International Planning Studies*, vol 14, no 4, pp425–435

Sonnino, R. (2010) 'Escaping the local trap: Insights on re-localization from school food reform', *Journal of Environmental Policy & Planning*, vol 12, no 1, pp23–40

Sukkel, W., Stilma, E. and Jansma, J. (2010) *Verkenning van de milieueffecten van lokale productie en distributie van voedsel in Almere*, Praktijkonderzoek Plant & Omgeving, Lelystad

Terragni, L., Torjusen, H. and Vitterso, G. (2009) 'The dynamics of alternative food consumption: Contexts, opportunities and transformations', *Anthropology of Food (Online)*, vol S5, http://aof.revues.org/index6400.html (accessed 28 June 2006)

Thompson, C. and Coskuner-Balli, G. (2007) 'Enchanting ethical consumerism', *Journal of Consumer Culture*, vol 7, no 3, pp275–303

USDA (United States Department of Agriculture) (2009) *2007 Census of Agriculture*, USDA, Washington, DC

Van der Ploeg, J. D., Renting, H., Brunori, G., Knickel, K., Mannion, J., Marsden, T., De Roest, K., Sevilla-Guzmán, E. and Ventura E. (2000) 'Rural development: From practices and policies towards theory', *Sociologia Ruralis*, vol 40, no 4, pp391–408

Watkins, K. (2002) 'Is Oxfam right to insist that increased access to Northern markets is a solution to the Third World's problems?', *The Ecologist*, vol 32, no 6, p34

Watts, D., Ilbery, B. and Maye, D. (2005) 'Making reconnections in agro-food geography: Alternative systems of food provision', *Progress in Human Geography*, vol 29, no 1, pp22–40

Winter, M. (2003) 'Embeddedness, the new food economy and defensive localism', *Journal of Rural Studies*, vol 19, no 1, pp23–32

Wiskerke, J. (2009) 'On places lost and places regained: Reflections on the alternative food geography and sustainable regional development', *International Planning Studies*, vol 14, no 4, pp369–387

Zezza, A. and Tasciotti, L. (2010) 'Urban agriculture, poverty, and food security: Empirical evidence from a sample of developing countries', *Food Policy*, vol 35, no 4, pp265–273

# 7

# Fair Trade: Buying and Selling Consumer Trust

This chapter aims to:

- explain fair trade as a strategy to contribute to sustainability in global food trade by introducing its history, objectives, institutions and practices;
- present fair trade coffee as an example;
- discuss the successes and limits of fair trade.

## Introduction

Only 50 years passed from the time when volunteers in the US began selling handicrafts produced by Mexican poor in the 1940s to the development of professional fair trade-labelling organizations involved in worldwide fair trade production and commercialization worth over €3 billion. Fair trade has transformed from a marginal initiative into an important alternative standard in global food trade. As both a standard and a certification scheme, the explicit intent of fair trade is to contribute to more equitable or 'fair' relationships in the global food market. Smallholder producers in developing countries are supported to increase their share of the commodity price and secure their position in the marketing channel.

How these aims are realized in practice and whether they actually result in sustainable improvements for poor producers in developing countries are questions for further analysis. Therefore, this chapter starts by introducing fair trade and its formal aspects, such as the labelling arrangements and relevant institutions (Fair Trade Labelling Organizations International, certifiers, etc.). Next, we illustrate this with the case of coffee, which was the first and is still one of the most important fair trade food products. Further illustration of the significance of fair trade is provided through presenting the sales of other fair trade products and by discussing the impact of fair trade on both producers and consumers. Finally, we discuss some critical comments on the aims and practices of fair trade and discuss its future perspectives as an innovative relationship between producer and consumer and examine whether this model can contribute to more sustainable global food provision.

## Historical Background

The first fair trade initiatives date from the 1940s and 1950s, when handicrafts produced by poor women in Mexico and Puerto Rico were sold by NGOs in the US and the UK. The first 'Fair Trade' shop to sell these handicrafts opened in the US in 1958. In the 1960s and 1970s, the initiative expanded through the creation of (Third) World shops in the Netherlands, Oxfam shops in the UK and similar initiatives in other countries. These shops started to sell food products from co-operatives in developing countries, in particular, cane sugar, coffee and tea, while still selling handicrafts. Ultimately, however, volunteers who ran the shops were less interested in selling products than in raising the shopping public's awareness of unjust global trade relationships and the need for change. These shops and their volunteers were part of a wider social movement that campaigned for 'trade not aid' and for more equity between the 'South' and the 'North' in international trade. Around the same time, alternative trade organizations developed an independent distribution channel to sell coffee (and other food products, such as sugar and honey) directly from producers in developing countries to consumers in Europe and the US, bypassing conventional market channels. In 1973, the first alternatively traded coffee produced in Guatemala was exported to the Netherlands. Alternative trade organizations did not want to engage with the multinational corporations that were active in food trade and retail because, as is said by many, 'they only exploit the poor farmers'. As a consequence of this choice, their initiative remained rather small scale and had hardly any significant impact on the global coffee market.

Despite all good intentions, this alternative trade did not significantly change the lives of many producers, so activists started looking for other ways to create more impact. They introduced the 'Max Havelaar' label in the Netherlands in 1988. The Max Havelaar Foundation was a certification initiative that intended to label coffee from existing brands as 'fair trade' and have these products available in regular distribution channels (including in supermarkets), so that many more consumers could easily buy them. If successful, this would rapidly expand the possibilities for poor farmers in developing countries to sell their produce for a guaranteed price to committed traders. Towards this aim, the Max Havelaar Foundation developed strict guidelines for producers, traders and roasters of coffee, and if these guidelines were followed, the coffee could be sold with the Max Havelaar label on the package through conventional market channels. This fair trade-labelling scheme was an attempt to mainstream the alternative market, and over time it has come to include many different food products[1] besides coffee: tea, bananas, other fresh and dry fruits, cocoa, wine, juices, fresh beans, honey, etc. The Max Havelaar initiative was rapidly followed in other European countries, and by the 1990s it was also adopted in the US and Japan. Initiatives in each country developed their own criteria and standards, but most countries called their label 'Fair Trade', because they were unfamiliar with the typically Dutch Max Havelaar.[2] The rapid spread of fair trade-labelling initiatives resulted in a substantial expansion of the market for poor producers to sell their products. Commercialization moved away from the isolated alternative trade circuits of World and Oxfam shops and towards conventional supermarkets; currently, over two thirds of fair trade-labelled products are sold through mainstream catering and retailing companies. By 2009, the fair trade retail sales topped $3.3 billion worldwide (FLO, 2010).

The next step in the institutionalization of fair trade was the establishment in 1997 of Fairtrade Labelling Organizations International (FLO), which was formed to coordinate the national initiatives that had previously operated independently. FLO defined worldwide standards for production and trade of fair trade products and thereby clearly contributed to further professionalization, harmonization and international collaboration within the global fair trade movement.

Recently, fair trade towns were added as a new strategy to promote fair trade, with Garstang in the UK as the first in 2005. Today there are over 400 fair trade towns in the UK, 44 in Belgium and others in Austria, Canada, Ireland, Italy, Finland, Norway, Sweden and many other countries.

Fair trade has become a well-established standard within international (food) trade, both at the producing end and the consuming end of the supply chains, and the initiative is increasingly institutionalized as well.

## Goals and Standards of Fair Trade

The fair trade movement aims to establish an international trade system that is based on fair conditions for farmers and workers in disadvantaged regions of the Third World. They define fair trade as:

> *a trading partnership, based on dialogue, transparency and respect, that seeks greater equity in international trade. It contributes to sustainable development by offering better trading conditions to, and securing the rights of, marginalized producers and workers – especially in the South. Fair Trade Organizations, backed by consumers, are engaged actively in supporting producers, awareness raising and in campaigning for changes in the rules and practice of conventional international trade.* (WFTO and FLO, 2009, p4)

Hence, organized producers (NGOs, co-ops) and sometimes certain plantations (see below) in developing countries are guaranteed a minimum price for their products, independent of the world market price. Moreover, direct and long-term partnerships are built between producers and importers, thereby providing a reliable basis for financial, technical and organizational capacity building. Consumer prices for fair trade products are higher than the prices of conventional ones because they are not simply 'market-conforming', but based on the notion of 'a fair return', covering the real production costs, supplemented with a social premium for developmental and environmental purposes. Initially, the criteria for fair trade food products focused on social and economic concerns, but later environmental considerations were included as well. This was in response to consumers, who assumed that fair trade food also had positive ecological impacts. Nevertheless, the criteria for producers' environmental performance remain purposefully rather vague in order not to exclude small-scale farmers, who may find it difficult to implement strict environmental guidelines. As the fair trade movement focuses on unfair trade relations between North and South, fair trade labelling is limited to Southern producers, hence small producers in the North cannot opt for it.

The main actors in the international fair trade networks are producers, importers, labelling organizations and sellers. Producers in developing countries are organized in democratic co-operatives or working for socially engaged companies. Fair trade importing organizations buy food from these producers and sell it via their own distribution channels or to conventional retailers. Conventional traders may also import fair trade food products, provided they are certified and subscribe to fair trade guidelines. Fair trade-labelling initiatives are operational in many countries. For instance, Max Havelaar started in the Netherlands in 1988, the Fairtrade Foundation in the UK in 1994 and Transfair in the US in 1999. Fair trade-labelling initiatives license the label 'FAIRTRADE Certification Mark' on products and promote fair trade in their respective territories, but they do not trade food or other products themselves. Fair trade products are sold via supermarkets, organic food shops, catering companies, (Third) World and Oxfam shops and local (church) groups, as well as directly from importers.

Several national fair trade-labelling initiatives established FLO in 1997 as their umbrella organization. Only labelling organizations were official members of FLO at first, but in 2007, regional producer networks from developing countries entered as well. Today the FLO Board consists of 14 members, composed of representatives from fair trade-labelling initiatives (5), representatives from fair trade-certified producer organizations (4; at least one from each of the regional producer networks), representatives from fair trade-certified traders (2) and external independent experts (3).[3] FLO sets fair trade standards, supports producers to gain fair trade certification and develop market opportunities, and coordinates global fair trade strategies to increase its impact and become more effective. The organization also promotes trade justice through participating in international debates on trade and development (see Box 7.1 on the Max Havelaar Foundation in the Netherlands, one of the founding members of FLO).

## Box 7.1 The Max Havelaar Foundation

The Max Havelaar Foundation was established in 1988, motivated by a call from Mexican coffee growers. They argued 'aid is good, but a fair price for our coffee is even better. Then we no longer have to beg for support.' If they could receive a fair price for their coffee beans, small-scale farmers would take their future into their own hands.

The Max Havelaar Foundation intended to change the trade system in support of producers from developing countries. To link producers with consumers, the Max Havelaar Foundation developed the Max Havelaar label for products produced and traded according to the fair trade standards. Many people think that Max Havelaar is a brand that markets its own products, but this is not the case. Max Havelaar is an independent label for fair trade and neither produces nor trades coffee or other food products. The core activities of the Max Havelaar Foundation are:

- offering independent certification via the Max Havelaar label for fair trade;
- convincing and facilitating the market to create supply and demand for fair trade products;
- collaborating with partners to support farmers' organizations and their members;
- increasing awareness in society about the need for fair trade and the importance of the Max Havelaar label for fair trade;
- searching for and realizing new product options in collaboration with licence holders;
- controlling companies that use the label on their products.

Today, the Max Havelaar label can be found on more than 250 different products in the Netherlands, ranging from peanut butter to towels.

*Source:* www.maxhavelaar.nl/ (accessed 21 October 2010)

Fair trade standards are defined by the FLO Standards Committee, whose members are appointed by the FLO Board. Its decisions must take into account the views of all relevant stakeholders and remain in line with FLO's overall mission and policy statements. The fair trade certification process itself is run by an independent company: FLO-CERT GmbH.[4] FLO-CERT assists producers by offering certification services to farmers and firms in more than 70 developing countries and by helping to foster long-term relationships and good practices with traders of certified fair trade products. The company independently audits both producers and traders who want to join the fair trade market and verifies their actual performance according to fair trade standards. Separate fair trade standards exist for producers (depending on their product) and traders. Certificates are issued only after a physical inspection has confirmed that all relevant fair trade standards have been complied with, and continued compliance is assured through a series of surveillance activities (see Box 7.2 for details on the certification process).

FLO-CERT insists that producers should make economically rational decisions because certification is an investment and once certified, producer organizations must pay for annual audits and may also need to invest in improvements to secure their certification for the future. In addition, being certified does not guarantee that producers can sell their products on the fair trade market, as they have to find buyers who are willing to purchase their products under fair trade terms. Selling fair trade products requires buyers to also be licensed with

## Box 7.2 The fair trade certification process

First, producers must decide which of the standards they want to qualify for, depending on the product and their organizational structure. Producers may be small farmer organizations, plantations, or factories.

The producer organization must apply to the company FLO-CERT for certification. If they qualify, FLO-CERT will inform them and request them to pay an application fee of €500 (about $700) and prepare for certification by complying with various criteria. Once all necessary documentation has been received and approved, a FLO-CERT certification manager carries out an on-site audit, which is paid for by the applying organization, to check them against the required compliance criteria.

These audits vary in length and intensity but always involve the following phases: an opening meeting with all stakeholders, site visits and interviews with representatives, members and non-members to cross-check information, and a closing meeting.

Then the certifier may highlight some issues that need to be addressed in order for the producer organization to comply with fair trade standards. Once compliance is determined, the FLO-CERT staff will issue a certificate, and the producer will be eligible to trade under fair trade terms.

*Source:* www.flo-cert.net/flo-cert/main.php?id=29 (accessed 23 October 2010)

the fair trade label from the fair trade-labelling initiative in the country where they want to sell their products. So ultimately, the entire supply chain that the product passes through, from producer to retailer, must be certified and audited to ensure the whole chain is compliant with fair trade standards.

The fair trade label is a voluntary private label, so producers are free to participate or not, and the initiative is independent from national and international governments. Nevertheless, fair trade producers and traders operate in the actual global market, and therefore, they have to respect existing international trade law. National governments may not actively support fair trade because WTO regulations for trade policy oblige them to be non-discriminatory in their interventions on trade. Measures from governments making fair trade products more attractive than non-fair trade ones would violate their obligation to refrain from imposing 'unjustified barriers to trade' (see also Chapter 4).

The FLO considered the rapid growth of fair trade sales in recent years (see below) as an incentive to review its own methods of operation. Organizational members argue that FLO needs to broaden, deepen and strengthen its operation so that fair trade can scale up in a way that is true to its core beliefs (FLO, 2009). FLO members also want to strengthen fair trade's position in the global market. The focus of this review is to become more responsive to producers' needs regarding the development and implementation of standards, as well as strategies for support and empowerment. Fair trade should not create artificially high barriers for producers or become a 'one size fits all' approach. 'Our empowerment model will provide a basis for producers to prepare their own development plans, including the milestones and indicators needed to monitor implementation' (FLO, 2009, p10). To this aim, FLO wants to orient its standards towards three different areas through a participatory process: business and development, production and trade.

## Fair trade, ethical trade and other labels

Fair trade is not the only sustainability label that is used in international food trade, and it is regularly mixed up with 'ethical trade', possibly leading to confusion among consumers. Sometimes fair trade is considered part of ethical trade, but in effect they are two different strategies. Both aim at ensuring that trade between 'South' and 'North' neither harms producer/worker welfare nor destroys the environment at production sites (Hughes, 2004). But fair trade is primarily aimed at changing the unequal terms of trade, while ethical trade intends to guarantee worker-friendly labour practices (Freidberg, 2003; Smith and Barrientos, 2005). Ethical trade tries to achieve this aim through adopting voluntary codes by large corporations, generally regrouped under the heading of 'corporate social responsibility' (Barrientos and Dolan, 2006). Several codes of conduct (or practice) are developed to support ethical trade, often on the basis of first-party declarations (CSR, Global Reporting Initiative (GRI)), but independent labels exist as well, such as SA 8000 and ISO 14000. In the world of food, companies such as Groupe Danone, Unilever, Nestlé, Starbucks and AHOLD have chosen this strategy. The principal difference in practice between fair and ethical trade is that fair trade guarantees poor producers in developing countries a minimum price for their products, while ethical trade offers no price guarantees.

Other sustainability labels, such as the Rainforest Alliance and Utz Certified, compete with fair trade labelling in the same market. The Rainforest Alliance (launched in 1987) aims to protect biodiversity and promote sustainability. To be certified, products such as coffee, tropical fruits, timber, cocoa and tea must be produced following standards developed by the Sustainable Agricultural Network (SAN). This label offers no price guarantees, but workers/producers have to receive at least a minimal wage, and their labour conditions must comply with International Labour Organization (ILO) conventions. The Rainforest Alliance label has become rather popular in the US for coffee and bananas. Another private label with growing importance, 'Utz Certified, Good Inside' (founded in 1997), aims to promote sustainability throughout global supply chains, with a focus on professionalizing them in their social, environmental and economic dimensions. In this third-party certification scheme, producers must show that they follow a code of practice based on GlobalGAP (Renard, 2005). There are no price guarantees, but by improving producer qualifications, product quality is expected to increase and thereby gain a stronger position on the market, with a higher price as a result. The main products are coffee, tea, cocoa, cotton and palm oil. Farmers consider Utz as an instrument for gaining market access, while retailers and roasters see it as a marketing tool to show their commitment to global sustainability (Muradian and Pelupessy, 2005). In the coffee sector, Utz has rapidly conquered important parts of the market in Western Europe and the US.[5] The Rainforest Alliance and Utz labels share many characteristics with fair trade, with the crucial exception of guaranteeing minimum prices for producers and paying a premium for community investments, as this element is unique to fair trade.

## The case of coffee

Coffee, today considered a primary food commodity, is enormously valuable to the economies of many developing countries. For some of the world's least-developed countries, exporting coffee accounts for the majority of their foreign-exchange earnings. Some of it is grown on large plantations, but most comes from farms with less than 10 hectares, involving over 25 million people in more than 50 countries. For many of them, coffee is a labour-intensive crop that frequently yields very little financial return. The principal coffee-producing countries are Brazil, Colombia and, more recently, Viet Nam. International coffee trade is dominated by a small and declining number of large trading companies based in the US and Europe.

The world market for coffee has always displayed high price volatility. This volatility prompted international governments to negotiate the first International Coffee Agreement in 1962 to stabilize the coffee market. Quotas were introduced to withhold excessive coffee supplies from the international market, while coffee consumption was promoted. Several subsequent International Coffee Agreements were reached as a result of coordination with the International Coffee Organization. Unfortunately, negotiations for the 1989 version of the International Coffee Agreement collapsed, causing coffee prices to drop to less than $0.80 per pound. This moment marked the end of governmental authorities' attempts to regulate the global coffee market, and more liberal market solutions were put forward to replace those earlier efforts.

Until the 1980s, national governments had played a vital role in supporting the production of coffee through credit-based input schemes, extension services, national systems of quality control and national pricing systems. But this started to change, and by 1990 the world of producing and trading coffee had definitely transformed. The producer cartel had collapsed, and the capacity of governments in developing countries to support their farmers and regulate the national coffee market had been greatly reduced because of structural-adjustment programmes imposed by the International Monetary Fund (IMF). Private contracts among producers, traders and industrial processors began to dominate the international coffee trade. Worldwide market coordination was reduced and therefore price instability increased. Despite several attempts, until today no successful global, government-based, coffee trade agreement has been established. In 1994, a new International Coffee Agreement was negotiated, but then it was decided that coffee prices would no longer be regulated. That same year, frost in Brazil threatened crops, and a pound of coffee escalated to a high of $2.80 per pound.[6] The international coffee market has moved away from a rather formal and stable system with active producer (i.e. country) participation to a more informal, inherently unstable and buyer-dominated one.

These transformations coincided with a growing differentiation in taste among consumers in importing countries. Until then, simple quality conventions, combining price with several crude physical properties of the crop, linked suppliers with these traders. But this simple matrix disappeared, and complex quality standards proliferated because the degree of differentiation in coffee blends and prices has grown significantly in the last decade (Ponte, 2002). This proliferation of grades and standards makes monitoring increasingly costly, and buyers are continuously trying to transfer these costs to producers (Gibbon, 2001), creating a permanent pressure on producers to adapt their cultivation practices and to incur the costs associated with certification.

The disappearance of global governance arrangements and the introduction of complex and diverse quality standards reinforced the existing asymmetrical balance of power in the coffee-supply network. Farmers in producing countries saw their position weaken, while importers, roasters and retailers were able to strengthen their positions (Fitter and Kaplinsky, 2001). The coffee price for farmers showed a declining trend, which was exacerbated by their decreasing share in the total market value of the crop. In the early 1990s, coffee-producing countries earned around $10–12 billion, while the value of the coffee sold by retailers was about $30 billion. In 2002, the value of the retail sales exceeded $70 billion, but the coffee-producing countries only received $5.5 billion (Osorio, 2002).

In a period that is now known as the 'Coffee Crisis' (October 2001), coffee prices fell almost overnight to an extreme low of $0.45 per pound. Hundreds of thousands of farmers were forced out of business, and over 100 million growers, processors, traders and retailers dependent on coffee were affected. This crisis resulted mostly from an overproduction of coffee. Vietnamese farmers were cultivating huge amounts of lower-quality Robusta coffee, and this expansion coincided with a drastic increase in Brazil's coffee production. This caused supply to far exceed demand. In 2002, 8 per cent more coffee was being produced than consumed, and much of it was low quality. The world market price of high-quality coffee was also dragged down by the rising supply of low-quality coffee.[7] This coffee crisis and the connected negative social and environmental impacts formed the driving force behind the introduction of fair trade coffee in the 1980s. Fair trade was intended to act as a safety net against the unpredictable world market, providing security to coffee producers, so that they get a price that covers their average costs of sustainable production. For this, FLO has developed the following fair trade standards for coffee:[8]

- Producers are organized in independent democratic organizations, and all members have an equal right to vote.
- Environmental standards restrict the use of agrochemicals and encourage sustainability.
- Producer organizations are guaranteed a minimum price (see Box 7.3 for details).
- A fair trade premium of $0.10/pound is added to the purchase price for social and economic investments at the community and co-operative levels.
- Pre-export lines of credit are available for the producer organizations (up to 60 per cent of the purchase price).

## Box 7.3 Fair trade coffee prices

Fair trade prices for coffee include the following considerations:

- Producer organizations are paid a floor price (Fair trade Minimum Price) of $1.25 per pound for fair trade, certified washed Arabica and $1.20 for unwashed Arabica, or the market price, if higher.
- For fair trade-certified organic coffee, an extra minimum differential of $0.20 per pound is being applied.
- A fair trade premium of $0.10 per pound is added to the purchase price and is used by producer organizations for social and economic investments at the community and organizational levels.

*Source:* www.fairtrade.net/coffee.html (accessed 23 October 2010)

Over the years, many producer organizations have acquired fair trade certification, and in 2009 more than 73,000 metric tonnes of certified coffee was sold to consumers. Still, it needs to be recognized that fair trade coffee represents only a very small share of the market: about 4 per cent of global coffee production.[9]

Fair trade's priority is on improving coffee producers' social conditions, but recently environmental consequences have begun to be included as well. Environmental impacts result from the intensification of coffee production that led to the increased use of pesticides, fungicides and artificial fertilizers. The introduction of monoculture coffee systems also meant aggravating soil erosion and deforestation, thereby endangering biodiversity, especially in vulnerable locations with high conservation value. Ecological sustainability was integrated in fair trade labelling at a later stage, but presently integrated crop management and specific guarantees for shade-grown production are required for certification (Abbott et al, 1999). Many European consumers value stricter environmental performance, however, and consequently, a substantial part of fair trade-labelled coffee is also produced and certified according to organic standards.

## Fair Trade Food Sales and Consumers

The fair trade movement's choice to opt for selling fair trade-labelled products through the conventional market system means their promoters have to collaborate with private companies, such as traders and processors and large retailers. Nowadays, fair trade-labelled food products are sold through supermarkets, next to the independent distribution channels of specialized fair trade shops, church and community groups and direct sales. In particular, the inclusion of fair trade products in conventional supermarkets has generated a formidable growth in sales in recent years (see Table 7.1).

As mentioned, fair trade standards differ for each type of producer organization, but each food product also has its own particular requirements. For instance, in 1996 bananas were the first fresh-fruit product that came under the

**Table 7.1** *Fair trade sales volumes*

| *Product* | *2004* | *2009* | *Growth (%)* |
|---|---|---|---|
| *Tea* | 1964 | 11,524 | 487 |
| *Coffee* | 24,222 | 73,781 | 205 |
| *Bananas* | 80,641 | 311,465 | 286 |
| *Fresh fruit* | 5157 | 20,091 | 290 |
| *Sugar* | 1961 | 89,628 | 4471 |
| *Honey* | 1240 | 2065 | 66 |
| *Juices* | 4543 | 45,582 | 903 |
| *Rice* | 1383 | 5052 | 265 |
| *Cocoa* | 4201 | 13,898 | 231 |
| *Wine** | 618 | 11,908 | 1827 |
| *Dried fruit* | 238 | 541 | 127 |

*Note:* All products in million tonnes, except (*) in 1000 litres.

*Source:* FLO (2006, 2010)

fair trade label and they rapidly reached a market penetration of up to 20 per cent in Switzerland (Roozen and Van der Hoff, 2001). Spreading this success to other countries was challenging because applying fair trade principles to a highly perishable product in a competitive market was difficult. The logistics and marketing of bananas had to rely more on conventional methods than was the case for products such as coffee or cocoa, which could be preserved more easily (Shreck, 2008). Therefore, unless a substantial market share was reached in a country, fair trade-labelled fresh food products remained dependent on conventional distribution systems and their willingness to give priority to fair trade products.

Coffee, cocoa, bananas and other fair trade-labelled food products have been on the market in several Western European countries since the 1980s and 1990s, but as Table 7.2 indicates, in recent years, several non-European countries have also had high growth rates, making fair trade food a global phenomenon. The market shares for fair trade products have been growing quickly since the early 2000s, whereby Switzerland and the UK have relatively high shares and Japan has remarkably low figures. Overall, the growth of fair trade sales has been impressive, reaching well beyond €3 billion in 2009. This figure was realized despite the economic crisis in 2008/2009, which probably reduced consumers' willingness to pay the additional price for fair trade products.

Buying fair trade food products should be more than simply buying a niche or particular brand product, but, as the Dutch importing organization Fairtrade Original suggests on its website, consumers should 'change the world shopping'.[10] Consumers are invited to become part of a social movement that aims to improve the lives of the people who actually produce their food. Through labelling, food acquires specific meanings beyond its specific material characteristics (see Box 7.4 for an example related to chocolate).

Fair trade standards and certification procedures are based on consumer concerns about farmers' livelihoods in developing countries. Genuine fair trade

Table 7.2 *Estimated retail value of fair trade sales (million €)*

| *Country* | *2004* | *2007* | *2009* |
|---|---|---|---|
| *Austria* | 15.8 | 52.8 | 72.0 |
| *Canada* | 17.6 | 79.6 | 202.0 |
| *France* | 69.7 | 210.0 | 287.7 |
| *Germany* | 57.5 | 141.7 | 267.5 |
| *Italy* | 25.0 | 39.0 | 43.4 |
| *Japan* | 2.5 | 6.2 | 11.3 |
| *Netherlands* | 35.0 | 47.5 | 85.8 |
| *Switzerland* | 136.0 | 158.1 | 180.2 |
| *UK* | 205.6 | 704.3 | 897.3 |
| *US* | 214.6 | 730.8 | 851.4 |
| **TOTAL (+ other countries)** | **831.5** | **2381.0** | **3394.2** |

*Source:* FLO (2006, 2008, 2010)

is defined at the production stage, where producers are considered to fall victim to declining prices and deteriorating opportunities for selling their produce on the global market. The key values presented through fair trade are solidarity between concerned consumers in the rich countries and poor farmers in developing countries to impose fairness as an essential principle in well-functioning global food markets. In contrast to organic labelling, fair trade also includes traders and importing firms in their certification schemes because they have to make sure that the correct prices are paid to producers and the agreed-upon trading conditions are implemented. Through fair trade, food-trading practices are changing because market transactions are no longer determined only by commercial norms and principles, but also by social and ethical principles. Fair trade prices are based on the notion of a fair return, covering production costs in addition to a social premium for development purposes (see Box 7.5).

Fair trade builds on domestic and civic norms, values and mentalities related to the responsibilities of global citizens. The FLO requirement that individual producers must be organized in co-operatives also strengthens civic norms and values among them. Consumers buying fair trade products are provided with 'personal relationships with farmers (through images, publicity, educational

## Box 7.4 UK retailer Co-op chooses fair trade cocoa for all its chocolate

In 2002, the Co-op decided to process all its own-brand chocolate bars from fair trade cocoa produced by small-scale farmers in Ghana. The retailer's head of branding explained that the company had chosen to focus on chocolate for a campaign to make trade fairer because of 'the obscene contrast between the pleasure derived from eating it and the suffering that often goes into supplying its main ingredient'. Fair trade pays a guaranteed minimum price and a premium to be used for community projects to help villagers build wells for clean water, schools and health centres.

*Source: Guardian* (2002)

## Box 7.5 FLO: Fair trade raises cocoa prices

As the global cocoa industry faces dwindling supply after years of underinvestment in cocoa farms in developing countries, FLO has announced higher prices for fair trade cocoa. 'Many cocoa farmers around the world don't have enough money for their families' basic needs – let alone to invest in new younger cocoa trees which can be their solution to a sustainable income from their cocoa farms for decades to come', says Rob Cameron, FLO CEO. 'With this higher fair trade income, cocoa farmers will get much needed funds to invest in farm improvements and their children's education'. FLO expects that cocoa farmers' organizations will earn at least $10 million in 2011 for development projects. In the past year, the number of cocoa farmer organizations certified by fair trade nearly doubled, and sales of fair trade cocoa grew by 35 per cent in 2009.

*Source: Biofach* (2010)

materials), trust and security in socially-responsible value claims and the elusive feel good factor' (Raynolds, 2002, p415). The initiative thereby intends to build innovative, direct social relationships between food producers and consumers beyond their economic relationship of buyer and seller, to create 'nourishing networks' (Whatmore and Thorne, 1997). The specific characteristics of such 'personal relationships' are emphasized, because not private, profit-oriented corporations but NGOs are controlling the principles and practices of the fair trade label. NGOs are expected to act more in the interests of smallholder farms in developing countries than commercial firms. Because all actors involved in fair trade throughout the supply network must be certified and follow strict guidelines, products in impersonal markets that are otherwise similar are differentiated, thus encouraging consumers to consider such differences when purchasing food. Fair trade consumption builds on a combination of life politics, the conscious choice for a particular lifestyle, and emancipator politics, aimed at addressing the global inequitable distribution of power and resources, often at a material cost to consumers (Levi and Linton, 2003; Lyon, 2006). The particular identity of fair trade commodities is '*fundamental to* and *necessary for* the creation of linkages between Southern rural livelihood struggles and morally reflexive consumers' (Goodman, 2004, p908). So, although purchasing fair trade products seems to be an individual buying choice, it becomes a political action as well when these choices are integrated into collective action, which is intended to demonstrate that the law of supply and demand and the domination of transnational companies are not inevitable (Renard, 2003). Such consumers adhere to a body of collective principles, and the act of buying fair trade products creates social ties among them.

At the same time, fair trade-labelled food products remain part of the conventional global food-trade system, and therefore, official regulations on food quality and safety are rigorously applied and monitored inflexibly. Delivery schedules and purchasing contracts are based on accepted industrial standards, which may thereby challenge fair trade producers and their organizations to adapt to these conventional modes of operation. In general, fair trade supply networks are supposed to adhere to normal commercial principles and have to function without subsidies or other forms of market regulation.

Fair trade is 'in the market but not of it', as Taylor (2005) argues. Maintaining this complicated position requires trust from various stakeholders, especially from consumers, as they have to pay the extra price. Fair trade-labelling initiatives try to convince consumers that labelled products are really contributing to improving the livelihoods of producers in developing countries. That certification is in the hands of independent, professional certifying agents, is an additional attempt to create and maintain consumer trust in the fair trade label.

## The Impact of Fair Trade

In a review of existing scientific studies on the impact of fair trade, Nelson and Pound (2009)[11] considered the economic, environmental and social dimensions of the innovative governance arrangement. There was strong evidence that fair trade makes a positive contribution to the economic position of smallholder farming families because they received higher returns and more stable incomes. There was also strong evidence that non-income impacts were at least as important as income benefits. Nevertheless, fair trade alone could not enable producers to escape poverty, as most of them are still merely surviving; fair trade needs to be supplemented with changes in development policies and coordinated with other actors and initiatives to be more effective. Also, Utting-Chamorro (2005) confirms that many problems (local financial services, rural–urban migration, etc.) remain, despite fair trade's positive effects on the livelihoods of small producers (see Box 7.6 for an example).

In a systematic review, Ruben (2008) compared the impact of fair trade on the lives of farmers with those not involved in fair trade. Ruben found different impacts, whereby fair trade led to larger claims on labour and to more speciali-

### Box 7.6 Impacts on a fair trade co-operative in Chiapas, Mexico

The extent to which producers benefit economically from alternative trade varies considerably and depends on the price of coffee in the conventional, non-fair trade market. When the price in this market rises above the fair trade minimum, the monetary benefits for the co-operative's members are relatively small. When the regular coffee price falls below the fair trade minimum, however, benefits to producers are much more substantial. In addition to the economic benefits in Chiapas, Mexico, there are non-economic ones. Co-operative ownership of processing and marketing facilities has enabled producers to engage in democratic participation. The survival of small-scale family farms has also kept ownership of means of production (including the land) in the hands of those actually engaged in the production itself. Environmentally, the vast majority of co-operative members is committed to shade-grown and organic-production practices, which have resulted in a much more environmentally benign system than the sun-grown, energy-intensive coffee production model that is encouraged by the Mexican state.

*Source:* Hudson and Hudson (2003, p422)

zation compared with non-fair trade. The impacts on household expenditures were found not to be significant, but fair trade definitely had positive results on crop yield (particularly for producers involved in organic production), production and price. These impacts led to much better access to credit and a greater willingness to take risks. The study also concludes that fair trade's focus has been on strengthening upstream linkages with producers' organizations, but very little attention has been given to downstream cooperation through 'co-ownership and profit sharing in processing and distribution networks' (Ruben, 2008, p45).

Overall, the environmental dimension has been studied insufficiently to make substantive claims about the impact from fair trade. On the social dimension, producers have profited from fair trade by increasing their self-confidence and strengthening their organizations (Utting-Chamorro, 2005). There is only limited evidence on the impact of fair trade on the position of workers, and there is no sufficient basis to judge whether and how much fair trade has effectively challenged gender norms and empowered women.

Therefore, despite its focus on the economic benefits of fair trade, it seems there are other benefits as well, although in some domains the evidence remains scarce (Taylor et al, 2005). Despite its good intentions, the direct and indirect benefits from fair trade do not seem to be spread evenly among all stakeholders, as observed by Tallontire (2000). According to a study on the collaboration between a UK-based fair trade coffee importer and a Tanzania-based coffee marketing co-operative, the importer was much more concerned about the ethical and developmental aspects of the relationship, while the producer was more interested in the market commitments.

Consumers buying fair trade products often do this with a broad agenda of supporting people in developing countries (Browne et al, 2000). Despite the higher price paid for fair trade products, Clarke et al (2007) conclude that it is not simply income level that defines an active engagement with fair trade, but rather, a complex assemblage of expertise, associational life and social capital (through churches, schools and trade unions). In a study on Belgium's consumers and their fair trade-buying behaviour, De Pelsmacker and Janssens (2007) found that better knowledge of fair trade and a positive attitude in general towards its issues had substantial effects on consumers' buying behaviour, while shopping convenience and price proved much less relevant. 'The most important hurdle for fair trade organizations is to fight indifference of people towards fair trade products' (De Pelsmacker and Janssens, 2007, p375). Consumers buying fair trade-labelled products are driven by social concerns, while the additional price they have to pay is of limited interest.

Overall, most studies confirm that fair trade has (limited) positive impacts for producers in developing countries, but that much depends on particular global market conditions and specific local circumstances.

## Challenging Fair Trade

Fair trade aims and their implementation have also attracted critical comments from different corners. The main debates circle around two issues: first, whether the transformative capacity of fair trade is too limited to make a real difference in the lives of poor producers and, second, whether small producers have sufficient influence in developing and implementing fair trade standards. These issues are further discussed in the next sections, but first we point to a criticism expressed by some economists, who argue on the basis of economic theory that the fair trade approach cannot be correct in principle. They argue that low food prices are the result of overproduction, so by offering artificially higher prices, fair trade only perpetuates and aggravates this problem because it prevents the effective functioning of price signals. Moreover, by guaranteeing a minimum price, fair trade eliminates incentives for producers to improve the quality of their products and increase their productivity (*Economist*, 2006). In a similar vein, Parrish et al (2005) argue that the strength of fair trade is rather on supporting demand-side market creation than on increasing supply-side production efficiency. As a consequence, less-efficient producers may remain in the market because they are supported by fair trade, while they should have shifted to other activities if market signals were allowed to be effective. Others consider this to be a much too narrow economic perspective and argue that successful empowerment of smallholders (Taylor et al, 2005) is as important as allowing the market to function, particularly when imbalances of power prevent these markets from functioning in a fair manner. These scholars are more interested in whether the fair trade approach is effectively transforming these unfair market mechanisms.

### The transformative capacity of fair trade

Fair trade's original intention was to change the imbalances between North and South on the global food market for the benefit of poor producers. This aim will not be achieved either when fair trade remains a marginal niche market, or when fair trade adopts less strict standards in order to be incorporated in the mainstream market.

If fair trade remains a niche market, the positive impacts from the fair trade regime will be limited to those farmers and workers who are included, while those excluded will be forced to work under unfair trade relationships. Currently, still only a small proportion of producers is included in fair trade arrangements, and these are often not the poorest in their communities (Wright, 2009; Neilson and Pritchard, 2010). Although fair trade labelling necessarily excludes producers who cannot comply with its standards (Goodman, 2004), this group of fair trade producers could be extended by improving the schooling and resources of the poorest farmers. In the global North, fair trade also risks remaining a niche market for alternative agri-food products. As Bryant and Goodman (2004) argue, the transformative possibilities of fair trade only go so far; the fair trade premium should not discourage consumers from buying the product. Next to this economic argument, Goodman (2004) suggests that consumers who are not in the position (in terms of class, education or knowledge) to be reflexive in their buying behaviour are also excluded. In response, (Raynolds, 2000) argues that fair trade should not target individual consumers with discretionary incomes making positive purchasing decisions, but enter the realm of citizen politics, where people make positive collective decisions about the nature of acceptable production and trade practices. Realizing this requires two-way networks of communication and exchange that span the North/South divide.

In contrast, if market shares increase, as the fair trade movement intends to achieve, some fear that the movement may reduce its alternative edge and dilute its original principles (Low and Davenport, 2005; Renard, 2005; Wright, 2009). Some are even witnessing a 'corporatization' (Jaffee, 2010) of alternative agri-food networks, where large commercial players take advantage of the profits offered by the niches of alternative food and the integrity they represent to consumers, while at the same time working to neutralize the transformative power of the standards underpinning that integrity. They consider the marketing of fair trade coffee by Nestlé, the sale of fair trade bananas by Sainsbury's and the partnership with fair trade organizations that Starbucks entered in order to double its purchases of fair trade coffee (Jaffee, 2010) all to be examples of fair trade being eaten away by big business. These critics fear that private companies that have shown interest in selling fair trade products will also seek to influence fair trade's standards and arrangements, formally or informally (Taylor, 2001).

To prevent this dilution of fair trade principles through conventionalization, some opt for further official institutionalization and enforcement by public authorities of fair trade standards imposed on all companies worldwide to improve the lives of all smallholders (Levi and Linton, 2003). Relying on voluntary arrangements would necessarily lead to the market reabsorbing the fair trade criteria and symbols (Renard, 2003). Others, however, put more trust in the social-movement character of fair trade, as the non-state nature of their standards offers much better prospects for societal influencing than those codified in law (Taylor, 2005; Jaffee, 2010). They consider the tensions associated with adopting fair trade's alternative approach while still operating in the conventional market as potentially constructive because activists in the fair trade movement are permanently challenged to protect its alternative logic. One concrete suggestion to strengthen this alternative character is to involve the producers and their organizations more actively in commercializing fair trade products (Raynolds and Ngcwangu, 2010). This would also reduce the (social) distance between food producer and consumer.

## The role of producers in developing fair trade standards

A final critical issue concerns the limited influence producers have on fair trade standards and marketing arrangements. Fair trade standards were defined by fair trade-labelling organizations in the importing countries, and FLO has taken a dominant role in their uniform spread across the globe. Also, the North has the power to dictate the type, quality, timing and volume of the product accepted for sale, and producers in the South can only comply. Shreck (2008) argues that the old colonial relations are continuing in a new guise and that reliance on the market as an engine of change perpetuates the structural imbalances that exist within it. It proves problematic to operationalize the ideals of fairness and equal exchange into concrete practices on the ground (Getz and Shreck, 2006). Fair trade standards are considered rigid, without much flexibility for accommodating variations in producers' circumstances. And, when going through the formal certification process, producers are also confronted with high costs (both time and money).

In response, as mentioned before, FLO is trying to involve producer organizations more explicitly in its internal procedures and in setting standards that should also be accessible for small-scale producers who have little resources at their disposal.

## Conclusion

Fair trade is an interesting attempt to contribute to more sustainability in globalized food provision because it tries to build connections between producers and consumers that are broader than the conventional buyer–seller relationship. It is an innovative governance arrangement that establishes worldwide links to maintain the global character of the flow of food, but introduces concrete sustainability improvements. Fair trade explicitly addresses the tension between conventionally operating market partners and mechanisms and the demand for more social justice and environmental change. In this way, fair trade builds on the understanding that consumers relate to food products not only in terms of nutritional and economic values, but also through values of equity and sustainability (Renard, 2003). Proponents from both the North and the South engage with these different values to foster a productive relationship (Jaffee et al, 2004). Contributions from civil actors, private firms and national and multilateral institutions are combined to organize more equitable global food trade.

Fair trade's success in terms of its growing market share opens up new challenges. As a response to consumer concerns about the social and environmental impacts of global food trade, maintaining consumer trust is a permanent challenge for organizations and companies involved in fair trade. The social marketing of fair trade, 'asking the consumer to buy into the idea of a fairer world', will become even more important in the future, as not only selected social categories are to be included, but also more mainstream consumers will be added if larger market shares are to be acquired.

Another challenge is how fair trade can better allow poor producers to profit from these improved market opportunities and grow out of poverty. This requires adopting a critical review of its internal standards and procedures in connection with the (price and quality) developments in conventional markets. But it also means that fair trade must consider its position in relation to other certification schemes, such as Utz, that private companies are rapidly adopting. Will fair trade continue to be regarded as the only real alternative to the conventional food market or will it become the basis of a pressure group that intends to change the ways in which the global food market operates?

---

*Take-home lessons*

- Fair trade is innovative because it uses the global food market to achieve social and environmental goals.
- Fair trade standards focus on social improvements but increasingly relate to environmental impacts as well.
- Becoming mainstream is fair trade's aim, but the movement is challenged on how to achieve this while maintaining its alternative, transformative character as a social movement.

---

## Notes

1 Fair trade covers food as well as non-food products, such as cotton clothing and handicrafts, but the presentation and discussion in this chapter are restricted to food products.
2 *Max Havelaar* is the name of a famous Dutch book, written in 1860, about the coffee trade between the Netherlands and the Dutch Indies (today Indonesia), accusing the coffee traders of making excessive profits because of the unequal colonial relationships that existed between the two countries.
3 See www.fairtrade.net/home.html (accessed 21 October 2010).
4 See www.flo-cert.net/flo-cert/main.php?lg=en (accessed 21 October 2010).
5 For instance, in the Netherlands the share of sustainably labelled coffee was estimated at 45 per cent in 2010, composed of Utz 39 per cent, fair trade 3 per cent, Rainforest Alliance 2 per cent and organic 1 per cent (while a third of fair trade coffee was also labelled as organic) (Oxfam-Novib, 2010).
6 See www.ico.org/history.asp (accessed 20 October 2010).
7 See www.fairtrade.net/coffee.html (accessed 23 October 2010).
8 See Note 7.
9 In 2010, the global coffee market experienced a steep increase in the price of coffee, with for January 2011 the highest coffee price since 1994. This price surge means that selling under fair trade conditions has become less attractive because the difference in price has become relatively small (ICO, 2011).
10 See http://fairtrade.nl (accessed 31 October 2010).
11 This study was commissioned by the Fairtrade Foundation, but the findings are reported independently.

## Further Reading

Raynolds, L. and Murray, D. (2007) *Fair Trade: The Challenges of Transforming Globalization*, Routledge, London: provides an overview of the background history and current practices of fair trade initiatives.

Ruben, R. (ed) (2008) *The Impact of Fair Trade*, Wageningen Academic Publishers, Wageningen: offers a collection of empirical findings on the impact of fair trade.

## References

Abbott, J., Roberts, S. and Robins, N. (1999) *Who Benefits?*, IIED, London

Barrientos, S. and Dolan, C. (eds) (2006) *Ethical Sourcing in the Global Food System*, Earthscan, London

*Biofach* (2010) 'Fairtrade raises cocoa prices', *Biofach Newsletter*, no 235, 29 October

Browne, A., Harris, P., Hofny-Collins, A., Pasiecznik, N. and Wallace, R. (2000) 'Organic production and ethical trade: Definition, practice and links', *Food Policy*, vol 25, pp69–89

Bryant, R. and Goodman, M. (2004) 'Consuming narratives: The political ecology of "alternative" consumption', *Transactions of the Institute of British Geographers*, vol 29, pp344–366

Clarke, N., Barnett, C., Cloke, P. and Malpass, A. (2007) 'The political rationalities of fair-trade consumption in the United Kingdom', *Politics & Society*, vol 35, no 4, pp583–607

De Pelsmacker, P. and Janssens, W. (2007) 'A model for fair trade buying behaviour: The role of perceived quantity and quality of information and of product-specific attitudes', *Journal of Business Ethics*, vol 75, pp361–380

*Economist* (2006) 'Voting with your trolley; Can you really change the world just by buying certain foods?', *The Economist*, 7 December

Fitter, R. and Kaplinsky, R. (2001) 'Who gains from product rents as the coffee market becomes more differentiated? A value-chain analysis', *IDS Bulletin*, vol 32, no 3, pp69–82

FLO (Fairtrade Labelling Organizations International) (2006) *Building Trust: Annual Report 2005–2006*, FLO, Bonn

FLO (2008) *An Inspiration for Change: Annual report 2007*, FLO, Bonn

FLO (2009) *Fairtrade's Labelling's Strategic Review*, FLO, Bonn

FLO (2010) *Growing Stronger Together: Annual Report 2009–10*, FLO, Bonn

Freidberg, S. (2003) *The Contradictions of Clean: Supermarkets Ethical Trade and African Horticulture*, IIED, London

Getz, C. and Shreck, A. (2006) 'What organic and Fair Trade labels do not tell us: Towards a place-based understanding of certification', *International Journal of Consumer Studies*, vol 30, no 5, pp490–501

Gibbon, P. (2001) 'Agro-commodity chains: An introduction', *IDS Bulletin*, vol 32, no 3, pp60–68

Goodman, M. (2004) 'Reading fair trade: Political ecological imaginary and the moral economy of fair trade foods', *Political Geography*, vol 23, no 7, pp891–915

*Guardian* (2002) 'Co-op to "double" fair trade chocolate sales', *The Guardian*, 27 November

Hudson, I. and Hudson, M. (2003) 'Removing the veil? Commodity fetishism, fair trade, and the environment', *Organization & Environment*, vol 16, no 4, pp413–430

Hughes, A. (2004) 'Accounting for ethical trade: Global commodity networks, virtualism and the audit economy', in A. Hughes and S. Reimer (eds) *Geographies of Commodity Chains*, Routledge, London and New York, pp215–232

ICO (International Coffee Organization) (2011) *Monthly Coffee Market Report; January 2011*, ICO, London

Jaffee, D. (2010) 'Fair trade standards, corporate participation, and social movement responses in the United States', *Journal of Business Ethics*, vol 92, pp267–285

Jaffee, D., Kloppenburg, J. and Monroy, M. (2004) 'Bringing the "moral charge" home: Fair trade within the North and within the South', *Rural Sociology*, vol 69, no 2, pp169–196

Levi, M. and Linton, A. (2003) 'Fair trade: A cup at a time?', *Politics & Society*, vol 31, no 3, pp407–432

Low, W. and Davenport, E. (2005) 'Postcards from the edge: Maintaining the "alternative" character of fair trade', *Sustainable Development*, vol 13, pp143–153

Lyon, S. (2006) 'Evaluating fair trade consumption: Politics, defetishization and producer participation', *International Journal of Consumer Studies*, vol 30, no 5, pp452–464

Muradian, R. and Pelupessy, W. (2005) 'Governing the coffee chain: The role of voluntary regulatory systems', *World Development*, vol 33, no 12, pp2029–2044

Neilson, J. and Pritchard, B. (2010) 'Fairness and ethicality in their place: The regional dynamics of fair trade and ethical sourcing agendas in the plantation districts of South India', *Environment and Planning A*, vol 42, pp1833–1851

Nelson, V. and Pound, B. (2009) *The Last Ten Years: A Comprehensive Review of the Literature on the Impact of Fairtrade*, Natural Resources Institute, London

Osorio, N. (2002) 'The global coffee crisis: A threat to sustainable development', paper presented at the WSSD, Johannesburg

Oxfam-Novib (2010) *Zuivere koffie. De Nederlandse supermarkten doorgelicht*, Oxfam-Novib, The Hague

Parrish, B., Luzadis, V. and Bentley, W. (2005) 'What Tanzania's coffee farmers can teach the world: A performance-based look at the fair trade–free trade debate', *Sustainable Development*, vol 13, pp177–189

Ponte, S. (2002) 'Standards, trade and equity: Lessons from the specialty coffee industry', *CDR Working Paper*, Centre for Development Research, Copenhagen

Raynolds, L. (2000) 'Re-embedding global agriculture: The international organic and fair trade movement', *Agriculture and Human Values*, vol 17, pp297–309

Raynolds, L. (2002) 'Consumer/producer links in Fair Trade coffee networks', *Sociologia Ruralis*, vol 42, no 4, pp404–424

Raynolds, L. and Ngcwangu, S. (2010) 'Fair Trade rooibos tea: Connecting South African producers and American consumer markets', *Geoforum*, vol 41, no 1, pp74–83

Renard, M. (2003) 'Fair trade: Quality, market and conventions', *Journal of Rural Studies*, vol 19, no 1, pp87–96

Renard, M. (2005) 'Quality certification, regulation and power in fair trade', *Journal of Rural Studies*, vol 21, no 4, pp419–431

Roozen, N. and Van der Hoff, F. (2001) *Fair Trade. Het verhaal achter Max Havelaar-koffie, Oké-bananen en Kuyichi-jeans*, Van Gennep, Amsterdam

Ruben, R. (ed) (2008) *The Impact of Fair Trade*, Wageningen Academic Publishers, Wageningen

Shreck, A. (2008) 'Resistance, redistribution, and power in the fair trade banana initiative', in W. Wright and G. Middendorf (eds) *The Fight over Food. Producers, Consumers, and Activists Challenge the Global Food System*, The Pennsylvania State University Press, University Park, pp121–144

Smith, S. and Barrientos, S. (2005) 'Fair trade and ethical trade: Are there moves towards convergence?', *Sustainable Development*, vol 13, pp190–198

Tallontire, A. (2000) 'Partnerships in fair trade: Reflections from a case study of Café direct', *Development in Practice*, vol 10, no 2, pp166–177

Taylor, E. (2001) 'HACCP in small companies: Benefit or burden', *Food Control*, vol 12, pp217–222

Taylor, P. (2005) 'In the market but not of it: Fair trade coffee and Forest Stewardship Council certification as market-based social change', *World Development*, vol 33, no 1, pp129–147

Taylor, P., Murray, D. and Raynolds, L. (2005) 'Keeping trade fair: Governance challenges in the fair trade coffee initiative', *Sustainable Development*, vol 13, pp109–208

Utting-Chamorro, K. (2005) 'Does fair trade make a difference? The case of small coffee producers in Nicaragua', *Development in Practice*, vol 15, no 3/4, pp584–599

WFTO and FLO (World Fair Trade Organization and Fairtrade Labelling Organizations International) (2009) *A Charter of Fair Trade Principles*, WFTO and FLO, Culemborg and Bonn

Whatmore, S. and Thorne, L. (1997) 'Nourishing networks: Alternative geographies of food', in D. Goodman and M. Watts (eds) *Globalizing Food: Agrarian Questions and Global Restructuring*, Routledge, London, pp287–304

Wright, C. (2009) 'Fairtrade food: Connecting producers and consumers', in D. Inglis and D. Gimlin (eds) *The Globalization of Food*, Berg, Oxford and New York, pp139–157

# 8

# Sustainable Fish Provision

The chapter aims to:

- present fish as an exemplary case of globalizing food provision;
- show how different public and private governance arrangements are introduced to reduce the unsustainability of fisheries and aquaculture;
- compare the legitimacy and effectiveness of various public and private governance tools.

## Introduction

Fish is an interesting case to illustrate the dynamics occurring in food provision in an era of globalization, as the place- and time-bounded activities of fishing and fisheries management are increasingly lifted out of their traditional local contexts and re-embedded into global networks and flows. Fish products such as salmon, pangasius and shrimp are sold the world over, while fishing vessels fish at locations far from the coasts of their countries of origin. Global fish trade involves multiple actors, including large private corporations and retailers, often operating at large distances from each other. At the same time, capturing and farming fish remains largely dependent on specific local ecological and climatic conditions, while fish consumption is still well-integrated into particular local eating habits and cultural norms. The realities of global fish trade and local fisheries are inextricably and irreversibly bound together through dynamic relationships in global flows of fish. At the same time, contemporary fish provision generates well-known sustainability impacts, such as depleting (or even collapsing) fish stocks and the use of antibiotics in aquaculture, so securing its sustainability has become an evident, although complicated, challenge.

This chapter starts by summarizing the present state of global fish provision, as it occurs through both capture fisheries and aquaculture. After introducing some further details on fisheries management and fish farming (aquaculture), we review various attempts to address sustainability challenges. First, nation state-based and internationally coordinated mandatory regulations (such as quotas, temporary closures, etc.) are discussed, with a focus on the role of the WTO. Next, we elaborate private governance arrangements, including certification and labelling schemes, such as Marine Stewardship Council (MSC), and consumer guides. In doing so, we focus on their respective understandings of sustainability and food and the roles of different stakeholders.

In the conclusion, we compare these different global fish-governance arrangements and assess their impact on sustainability in global fish production and consumption.

## Global Fish Provision

Fish is an essential source of protein in the diets of the world's population and is regularly consumed by billions of people (see Table 8.1). Fish contributes more than 35 per cent of the total animal protein supply of about 1.25 billion people in 39 countries worldwide, including 19 sub-Saharan countries (Tacon, 2010). Furthermore, some 44 million people depend (partly) on fishing and aquaculture for their livelihood, most of them through small-scale or artisanal activities in developing countries (Love, 2010).

Table 8.1 shows that although total fish production has increased substantially, the amount of captured marine fish has remained stable. The main species captured is anchoveta, an anchovy caught off the coasts of Peru and Chile that

**Table 8.1** *World fisheries and aquaculture production and utilization (million tonnes)*

| | *2002* | *2003* | *2004* | *2005* | *2006* |
|---|---|---|---|---|---|
| **PRODUCTION** | | | | | |
| INLAND | | | | | |
| Capture | 8.7 | 9.0 | 8.9 | 9.7 | 10.1 |
| Aquaculture | 24.0 | 25.5 | 27.8 | 29.6 | 31.6 |
| **Total inland** | **32.7** | **34.4** | **36.7** | **39.3** | **41.7** |
| MARINE | | | | | |
| Capture | 84.5 | 81.5 | 85.7 | 84.5 | 81.9 |
| Aquaculture | 16.4 | 17.2 | 18.1 | 18.9 | 20.1 |
| **Total marine** | **100.9** | **98.7** | **103.8** | **103.4** | **102.0** |
| Total capture | 93.2 | 90.5 | 94.6 | 94.2 | 92.0 |
| Total aquaculture | 40.4 | 42.7 | 45.9 | 48.5 | 51.7 |
| **Total world fisheries** | **133.6** | **133.2** | **140.5** | **142.7** | **143.6** |
| **UTILIZATION** | | | | | |
| Human consumption | 100.7 | 103.4 | 104.5 | 107.1 | 110.4 |
| Non-food uses | 32.9 | 29.8 | 36.0 | 35.6 | 33.3 |
| Population (*billions*) | 6.3 | 6.4 | 6.4 | 6.5 | 6.6 |
| Per capita food fish supply (*kg*) | 16.0 | 16.3 | 16.2 | 16.4 | 16.7 |

*Note:* Excluding aquatic plants.

*Source:* FAO (2009, p3)

## Box 8.1 Fishing Tonle Sap Lake in Cambodia

In Tonle Sap Lake in Cambodia, floating communities of full-time fishers try to make a living. They sell their produce to traditional and export markets. Traditional, domestic fish markets are characterized by limited managerial and entrepreneurial skills, resulting in low-quality and low-price fish (products). Export markets, by contrast, are increasingly controlled by efficient, capital-intensive, large-scale traders and trading associations. In both markets, fishers are marginalized by high informal taxes and gratuities paid to a range of government-sanctioned concessionaires. The very complex and diffuse trading system makes it very hard for small fishers to understand its dynamics and make use of its opportunities.

*Source:* Bush and Minh (2005)

is primarily used for fish meal as animal feed. Its catch in 2004 was 10.7 million tonnes, four times that of the second-most-captured species, Alaska pollock (Love, 2010), which is destined for human consumption. Since the mid-1980s, developing countries have surpassed developed countries in producing fish through expanding their fleets and intensifying their aquaculture systems. The creation of exclusive economic zones (EEZs) in 1977 contributed much to their growing importance because since then developing countries have been much better able to protect their natural wealth (Ahmed, 2006). Many developing countries also have substantial aquaculture activities and inland fisheries, each with its own specific characteristics (see Box 8.1).

Even though world fisheries' capture production has levelled off, international trade has continued to increase; its value grew from less than $10 billion in 1975 to some $86 billion in 2004 (FAO, 2006). Fish has become, in relative terms, one of the most important globally traded food products. Over 40 per cent of the world's fish production entered international trade in 2004.[1] Developing countries increased their share in these global fish exports from 37 per cent in 1976 to 50 per cent in 2001 (OECD, 2010), and thus fish has surpassed their traditional agricultural exports, such as coffee, sugar and rice. These countries mainly export high-value species, such as shrimp, lobster and tuna, but also low-value species such as pangasius.

Transnational companies are influential in organizing fish provision, although mostly in aquaculture, because capture fisheries are based on country-related fishing rights. These large companies are involved in vertical integration, from harvesting to marketing. See Table 8.2 for an overview of the largest corporations, confirming the important role that Asia, in particular Japan and China, plays in global fish production and trade.

The global consumption of fish is still growing, but capture fisheries can no longer produce sufficient quantities because overfishing has led to declining stocks. Indeed, since the 1950s, marine fish catches have increased dramatically when such technological innovations as onboard freezing systems and synthetic fibres in fishing nets were introduced. But, in the 1980s, the world suddenly became aware that these increasing fishing efforts had negative consequences when several stocks collapsed (e.g. Atlanto Scandinavian herring stocks

Table 8.2 *The ten largest seafood companies globally*

| | Company | Country of origin | Market capitalization (million US$) |
|---|---|---|---|
| 1 | Marubeni Corp | Japan | 8.450 |
| 2 | PAN Fish ASA | Norway | 3.004 |
| 3 | Nutreco Holding NV/Marine Harvest | The Netherlands | 2.278 |
| 4 | Toyo Suisan Kaisha Ltd | Japan | 1.728 |
| 5 | Nichirei Corp | Japan | 1.706 |
| 6 | Nippon Suisan Kaisha Ltd | Japan | 1.641 |
| 7 | Katokichi Co. Ltd | Japan | 1.324 |
| 8 | Cermaq | Norway | 1.288 |
| 9 | Austevoll Seafood ASA | Norway | 1.148 |
| 10 | Dalian Zhangzidao Fishery Group Co. Ltd | China | 1.146 |

*Source:* OECD (2010, pp113–114)

and cod stocks off Newfoundland) (OECD, 2010). In the 1970s and 1980s, the proportions of overexploited, depleted and recovering fish stocks increased substantially, but this situation has remained relatively stable over the last 15 years. Still, in 2007, about 28 per cent of the stocks were either overexploited (19 per cent), depleted (8 per cent) or recovering from depletion (1 per cent), while about 52 per cent of the stocks were fully exploited (FAO, 2009). Nearly all of the top-ten most-popular species, together accounting for about 30 per cent of the world's capture fisheries production, are fully exploited or overexploited. Overfishing reduces the existing stocks and their natural capacity for regeneration, which will inevitably result in lower catches over the longer term. Pollution caused by processing activities at sea, habitat degradation caused by trawling and the use of prohibited fishing methods further add to these problems. In addition, the destruction of coastal zones, wetlands and mangrove areas is impairing their function as natural spawning grounds and nurseries for replenishing marine fish stocks (Garcia et al, 1999). The FAO concludes that the maximum yield from all wild capture fisheries has probably been reached. Therefore improving fisheries management for recovering endangered fish stocks and conserving the remaining marine biodiversity is urgently needed, particularly for some highly migratory, straddling fishery resources, such as tuna, which pose complicated management challenges.

## Aquaculture

As the supply of fish from capture fisheries is stagnating, the production of fish from aquaculture has grown to meet the rising demand. Aquaculture is the farming of aquatic animals and plants in ponds or cages and has been the fastest-growing animal-food production sector globally, expanding at an average rate of 8.1 per cent annually since 1961. Food fish supply from aquaculture increased from 4.3 to 7.5kg/capita from 1995 to 2007 (Tacon, 2010). In 2007, 65.2 million tonnes (valued at $94.5 billion) were produced, largely in Asia (see Figure 8.1).

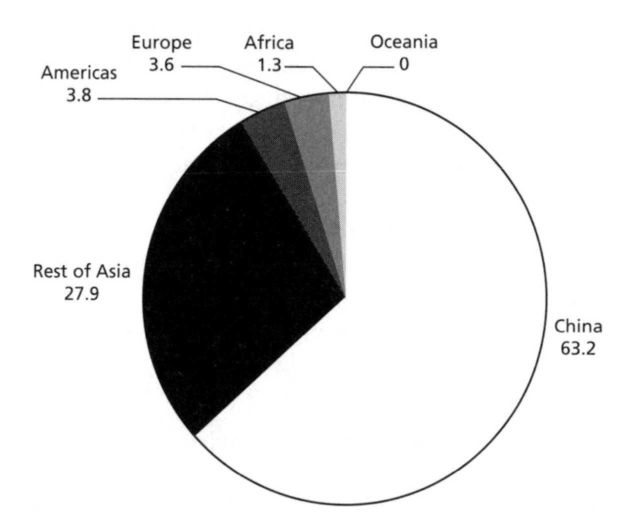


**Figure 8.1** *Distribution of aquaculture production (%, 2007)*

*Source:* Tacon (2010)

Most fish species produced from aquaculture are omnivorous or herbivorous, such as pangasius, tilapia and carp, which have relatively low environmental impacts. Some other species, such as salmon, however, are carnivorous and have a much higher environmental impact because they depend on capture fisheries for sourcing raw materials for feed formulation and seed inputs.

Aquaculture may have negative impacts, but much depend on the species, farming practice and country concerned (Tacon, 2010) (see Box 8.2). Environ-

## Box 8.2 Thai shrimp farming

Global shrimp production represents a value of $7 billion, or 20 per cent of the fish trade's total value. Leading countries in producing shrimp are China, India, Thailand and Indonesia, while Thailand is one of the largest exporters of fresh and frozen shrimp, with a value of €2 billion in 2010.

Producing shrimp entails several environmental risks, notably the destruction of mangrove forests and the routine use of antibiotics in the ponds. Hence the Thai government developed a regulation for shrimp production that has been in effect since 1991. The national authorities chose a command-and-control approach by directly controlling mangrove invasion and installing spatial planning to protect tourism, rice production and other activities competing with aquaculture. These efforts continued without the expected results because of non-compliance from producers, ineffective enforcement from local and regional authorities and global competition, which made unilateral measures risky, as Thai shrimp might lose out on the global market.

NGOs criticized the limited results in correcting the unsustainable practices in shrimp farming. In reaction, these NGOs either resorted to certification to promote sustainable shrimp farming or called for a boycott of farmed tropical shrimp. Also, producers organized in the Global Aquacultural Alliance initiated a code of conduct to rectify unsustainable practices. Over time, these initiatives, together with pressure from importers in Europe and the US, resulted in improved pond management practices.

*Source:* Sriwichailamphan (2007)

mental concerns include habitat loss from constructing ponds, pollution from the use of chemicals and excess feed, and escapes of farmed animals that may endanger wild stocks through diseases and genetic interaction. Resource-use concerns include feed selection, water, land and energy use and wild seed collection. Social concerns entail displacement of coastal communities; disruption of seafood prices, local food supply and food security; salinization of potable water and groundwater; social exclusion; and conflicts with tourism and recreational and commercial fishing. Potential food-safety concerns relate to possible contamination with heavy metals, pollutants, chemicals, medicines and pathogens. In sum, aquaculture may lead to many diverse problems, but most such problems can be mitigated and minimized by adequate production methods.

## Governing Fisheries

Conventional regulations on fisheries designed to reduce the problem of overfishing have existed for many years. These regulations concern the distribution among a selected number of fishers/vessels of individual rights (quotas) to fish a particular stock. This distribution is based on the understanding that overfishing is caused by competition between individual boats trying to catch the largest possible share of the stock, resulting in a classic case of 'tragedy of the commons' (see Chapter 3). By introducing an individual rights-based approach, the available sustainable catch in a fishery is shared among the fishing units, with the expectation that 'the race for fish' will end. If the sum of all individual shares is well-determined on the basis of the maximum sustainable yield calculated through adequate scientific measurement, this policy should prevent overfishing. When individuals have a secure share of the fishery, they no longer have an incentive to race, since this will not bring them more profit. Individual rights should be allowed to be transferred freely to others, for example, through an 'individual transferable quota', in order to maximize economic benefits.

Despite many fisheries having introduced these regulations some years ago, the pressure on the remaining fish stocks is not decreasing, particularly not on the high seas. Therefore, finding a more effective way to govern global fish production and trade and to increase its sustainability has become an urgent concern. In recent decades, multilateral agreements for fisheries management have been developed, but lately private market-based arrangements, such as certification and labelling schemes and consumer guides, have been established.

### Multilateral fisheries governance

Fisheries governance is complicated because of the open access nature of fishery resources. Fish stocks and fishing fleets straddle across national borders easily, and also, it is difficult to enforce agreements because fisheries have so many diverse characteristics.

National governments trying to impose strict regulations unilaterally on their domestic fisheries or on imported fish products face severe problems when securing their market position and maintaining good trade relations. The accepted principle of national sovereignty prevents government interference in other countries' waters or fisheries management systems. In response, a multilateral legal framework has been introduced to supplement existing national regulations and to create a global fisheries regime. Global governance arrangements essentially add another layer of governance to existing local and national governmental systems, with the purpose of regulating problems that cross national boundaries. This transnational governmental level becomes established through a transfer of power from local and national levels.

Global fishery governance started with the UN Convention on the Law of the Sea (UNCLOS) in 1982, while the 1992 UNCED underlined the need to create effective international fisheries management regimes. The 2002 UN World Summit on Sustainable Development (WSSD) in Johannesburg, South Africa, also produced an action plan for oceans governance. Under the auspices of the FAO and the UN, guidelines were formulated to protect existing fish stocks, and several regional fisheries management organizations (RFMOs) were established to manage high seas' and straddling fisheries. Presently, over 100 multilateral, regional and bilateral treaties exist to supplement the UNCLOS and existing national laws, including 14 UN agencies and 19 international governmental organizations that have the oceans as (part of) their responsibilities (Peterson, 1993). Despite their increasing number, however, the effects of these international agreements remain very modest because some governments do not seem very committed to implementing them, while others lack the capacity to put the guidelines and measures into practice. Overall, fisheries management has remained relatively low on governments' agendas and, in most countries, attracts little public attention beyond those directly involved (Allison, 2001). And, as direct governmental interference in international fish trade is restricted by WTO agreements, fish has become a globally traded food without a relevant, well-functioning global governance regime (Bush and Oosterveer, 2007).

## The WTO and fisheries management

Although fisheries are not explicitly covered by any particular agreement within the WTO, fisheries-related issues are included in several of its general agreements. These agreements are aimed at promoting free international trade but avoid putting limits on national sovereignty. Through multilateral negotiations, the WTO intends to create an 'even playing field' for competition between its member states and to reduce the presence of unjustified barriers to trade. Therefore, the WTO allows its member state governments to interfere with international trade only on the basis of product-related characteristics and not to regulate trade according to production methods (see Box 4.2 for the famous dolphin–tuna case). Hence, governments cannot distinguish between fish produced sustainably and unsustainably in its trade regulations.[2] Within the WTO, fish could be subject to the TBT and the SPS agreements (see Chapter 4) to preclude unnecessary barriers to international trade through technical regulations or public health and safety measures. However, except for a discussion between the EU and Peru on the definition of 'sardines', these particular agreements have had little relevance for international fisheries trade.[3]

Still, fish trade and management are explicitly covered within the WTO under its agreement on Subsidies and Countervailing Measures (SCM) (Ponte et al, 2007). This agreement has been on the Agenda of the Doha round since 2001, and negotiations are aimed at reducing or eliminating fisheries subsidies. Direct subsidies to fisheries are substantial and are considered to have a distorting effect on production costs and competitive positions in international trade.[4] Worldwide, annual fisheries subsidies are estimated to be about $34 billion, of which some $20 billion are considered as directly supporting overcapacity fishing. The biggest subsidizers are Japan ($5.3 billion), the EU ($3.3 billion) and China ($3.1 billion). Subsidies are seen to be responsible for overfishing because they encourage an excess capacity of fishing fleets (FAO, 2000; Potts and Haward, 2007). To strengthen discipline on fisheries subsidies, different proposals have been discussed in the WTO Negotiating Group on Rules. Some countries, organized in the 'Friends of Fish' group, proposed a general ban on all subsidies, with some limited exceptions. This group brings together countries with a fish-trade focus (New Zealand, Chile, Iceland) and countries with an environmental agenda, and it is supported by environmental NGOs, notably WWF and Friends of the Earth International (FoEI) (Ponte et al, 2007). This 'Friends of Fish' proposal claims that radical elimination of fisheries' subsidies is beneficial for trade, but also has additional value 'in terms of protecting the marine environment and also sustainable development, since so many countries depend on fishing as an important part of their economy' (US ambassador to the WTO quoted in ICTSD, 2007, p4). Tough fisheries subsidies discipline is considered an essential complement to strong fisheries-management programmes to ensure that wild fish stocks remain sustainable for future generations, because it slows down their overharvesting. By contrast, however, during the negations, countries such as Japan, Korea and Taiwan argued that there is no need for a general ban on fisheries subsidies, with the exception of those that are clearly problematic from an international competition point of view. Essentially, only those subsidies that result in an increase in existing fishing capacity should be prohibited (*Bridges Weekly*, 2007, 2008). So, first those subsidies that are specifically harmful should be identified and then banned, instead of simply implementing sweeping prohibitions. Developing countries argue that their artisanal and small-scale fishing industries have only a minimal impact on overfishing and overcapacity, and therefore, wide-ranging prohibitions on subsidies to protect the environment would be unduly punitive for them (Potts and Haward, 2007).

Thus, the WTO's involvement in the current international fishery trade and subsidies regime has both direct and indirect consequences on international fisheries governance. Direct consequences result through the implementation of its existing agreements and the debates about abolishing subsidies (Allison,

2001). Indirect consequences occur by excluding, or at least complicating, the implementation of trade regulations that explicitly include environmental aims. Essentially, the WTO approach to fishery governance is oriented to harmonizing the conditions for international trade and creating free markets, while environmentally beneficial effects are expected merely as by-products. When analysing attempts to bring in market discipline in managing the Alaska pollock fishery, however, Mansfield observes that this process created multiple market incentives, 'some of which may lead to environmental protection while others may not' (2006, p37). As these efforts at a multilateral level to increase sustainability in global fish provision and effectively manage existing fish stocks show only limited progress, other initiatives have been taken.

## Community-based fisheries management

Fisheries management based exclusively on formal biological science and general economic principles is increasingly questioned, as it has not prevented the further overexploitation of fish resources. This approach seems inherently unable to address present problems because of the way in which objectives are defined, limitations in the knowledge on which the objectives and manners are based and the top-down nature of their implementation. One alternative is to decentralize fisheries management and involve local communities through co-management, an arrangement where responsibilities for managing fisheries are shared between the government and local fishing communities (Hoof, 2010). Co-management is considered to be more democratic because of stakeholder participation and more effective because including practitioners' knowledge makes management measures more acceptable and applicable.

Community-based management also better allows for an ecosystem-based approach that implies managing a wider array of human activities than traditional fisheries management and gives attention to negative externalities resulting from the fishing activities themselves. A successful ecosystem-based fisheries-management process necessitates the involvement of actors from other sectors in addition to the fishing industry (Juda, 2002).

Some efforts have been made to realize this approach in practice, but these efforts have encountered several problems, including how coordination should be arranged among the different levels of governance and how conflicts within the local community itself could be solved.

## Fish labels

In recent years, several private labels and certification schemes have been introduced to promote sustainable fisheries.[5] These instruments seem to be increasingly preferred above quotas, moratoriums or other official governmental tools. Such private arrangements involve multiple actors in standard-setting and monitoring roles through networks established at multiple levels and aim at voluntary cooperation. Although environmental labelling dates back to the 1970s (Rubik and Frankl, 2005; Boström and Klintman, 2008; Parkes et al,

### Box 8.3 Saving dolphins through labelling

Two labels represent efforts to try to reduce the mortality of dolphins during the capture of tuna: 'dolphin safe tuna' and Agreement on the International Dolphin Conservation Program (AIDCP) labels. Both labels can be issued when tuna fishers avoid unintentionally killing dolphins.

The Earth Island Institute (EII) established an international monitoring programme in 1990 to monitor catches and shipments around the world and to guarantee companies and consumers that tuna caught in a 'dolphin-safe' manner is protecting dolphins. All fishing and carrier vessels; all processing, storage and transhipment facilities; and all procurement records related to the purchase, processing, storage, transport and sale of tuna must be made available for independent EII-approved monitoring. The organization collaborates with tuna companies – import associations, fishing fleets, canners and brokers – to establish 'dolphin-safe' policies for each company and is one of the largest private food-monitoring systems in the world.

The countries and regional economic integration organizations participating in the AIDCP announced a programme to certify and label tuna caught in the eastern Pacific Ocean consistent with AIDCP guidelines without killing or causing serious injury to dolphins. AIDCP dolphin-safe tuna certification is supported by a comprehensive and transparent multilateral tracking and verification system administered by member governments and the treaty organization to ensure full consumer confidence in the AIDCP dolphin-safe label.

*Source:* Accenture (2009); UNEP (2009)

2009), labelling fish is a relatively recent practice (Roheim and Sutinen, 2006). Among the first fish products to be labelled as sustainable was 'dolphin-safe tuna', introduced in the US in 1990 to guarantee consumers that their tuna was not caught while killing dolphins at the same time (see Box 8.3.) Several different private labels have been developed for capture fisheries and aquaculture (Accenture, 2009).

Sustainable international trade of fish requires market-oriented coordination connected to ecosystem-related dynamics and allows for increasing sustainability in fishing practices and management of fish stocks. Certification offers a coherent arrangement to promote sustainable fisheries throughout globalized supply chains. Certification confirms that a fishery operates according to a precise set of performance standards, so that a label can be awarded to products from that fishery, provided that the chain of custody is certified as well (Potts and Haward, 2007). According to the FAO (2005), a fishery's eco-label is a voluntary product label conveying environmental information to consumers, thereby seeking to create a market-based incentive for better management of fisheries. In recent years, the presence of eco-labelled seafood on the market has been increasing, and a broad spectrum of different labels is emerging. The range of different certification schemes reflects the variety of incentives to which multiple initiators and users are responding (Parkes et al, 2009). The fish-product label conveys information without the buyer having to comprehend all details, but to be successful, consumers must trust the labels.

Major fish labels are MSC and Friends of the Sea (FOS), but new initiatives are often introduced.[6] The presence of these different and diverse fish labels has led to calls for quality standards, so that retailers and consumers can make informed choices. Therefore, the FAO has developed guidelines for the eco-labelling of fish and fishery products from marine capture fisheries (FAO, 2005), which are considered the most comprehensive benchmark for fish labels. According to these guidelines, labels should be consistent with relevant international conventions, recognize the sovereign rights of states and comply with all applicable laws. In addition, they should be science-based, transparent, allow for fair participation by all interested parties, be non-discriminatory and not create unnecessary obstacles to entering international markets.

At present, the MSC label dominates sustainable fish labelling and this initiative is introduced in the next section. This is followed by a review of several other labels.

## *MSC*

The MSC belongs to the group of the most well-known and well-established sustainable fish labelling schemes. It was initiated in 1997 by Unilever and WWF to solve the global problem of overfishing. The organization evolved into an independent private organization in 1999, when the initiators withdrew (Gulbrandsen, 2009). MSC intends to use its eco-label and fishery certification programme 'to contribute to the health of the world's oceans by recognising and rewarding sustainable fishing practices, influencing the choices people make when buying seafood, and working with (its) partners to transform the seafood market to a sustainable basis'.[7] MSC wants to combine environmental, commercial and social interests to promote sustainable fisheries.

MSC standards apply to wild-capture fisheries only, to strengthen its credibility. The organization uses a science-based approach to identify non-visible product characteristics and determine a fishery's sustainability. The MSC label establishes a collaboration among all different social actors involved in capturing and trading (the products of) a particular fish stock (Gulbrandsen, 2009). The introduction and promotion of this scheme has been primarily driven by providers, such as fishers, processors and traders, while consumers were requested to buy sustainably labelled fish, thus contributing to a more sustainable food provision. The MSC label itself is exclusively environment-oriented because the organization contends that demanding higher social and economic performances than required through existing national legislation would bring sustainable fisheries into an even more unfavourable position vis-à-vis conventional fisheries. The economic and social benefits are seen as side effects because MSC certification allows producers to sell their catches to the higher end of the market. The MSC arrangement is innovative because it establishes previously unknown linkages between production and consumption and seeks to reconnect producers and consumers in the process of making fish-supply chains more sustainable. Consumer buying power is considered more effective when trying to influence producer practices than official regulations.

To reward environmentally responsible fisheries management, the MSC developed its Principles and Criteria for Sustainable Fishing as a standard for a third-party, independent and voluntary certification programme. Its specifications and allocation of functions set precise standards for production and processing methods and codify knowledge requirements. The development and promotion of these standards have been primarily provider-driven: NGOs, fishers and companies have been designing the criteria, procedures and (green) identities, which are subsequently 'offered' to citizen-consumers without further efforts to engage them in supporting more sustainable fishery practices or in learning more about their specific sustainability concerns in the case of fish. After an extensive certification process, a product label can be awarded for five years and subsequently renewed. The certifying and labelling procedure of a particular fishery is based on three general principles.[8] First, the fishery must not contribute to overfishing or depletion of the exploited fish population. Second, the fishery should allow for the maintenance of the ecosystems on which it depends. Finally, the fishery should have in place an effective management system that respects local, national and international laws and standards. All relevant stakeholders in a particular fishery are engaged in translating these general principles into a concrete and detailed management plan for the responsible and sustainable use of a specific fishery. This resulting management plan must be assessed by an independent certifier. In this plan, all actors concerned agree on how, how much and when fish should be caught, as well as on the implementation of accompanying measures when needed. All stakeholders have to participate in the process, but MSC considers support from environmental NGOs to be vital 'if our program is to offer industry the credibility they expect' (MSC, 2002, p1), so certifying agents should make efforts to contact them and request their participation in the process.

By December 2010, 102 fisheries had received the MSC label, all of which are located in OECD countries, except for 3 in South Africa, Viet Nam and Mexico, while 132 other fisheries were undergoing assessment.[9] When the products are sold via an MSC-certified chain of custody, consumers can buy fish with the blue MSC label, which guarantees its origin from a certified sustainable fishery. In supermarkets and specialized fish retail stores in industrialized countries, more than 5000 different MSC-labelled items from certified fisheries are available. The organization claims that in 2009 some 12 per cent of the world's total edible wild-capture fisheries, or over 7 million tonnes of seafood, were part of the MSC programme (MSC, 2009, 2010).

Criticisms regarding the MSC initiative include comments on the involvement of multinational companies, such as Unilever, together with globalized environmental NGOs, such as WWF, because they are seen as promoting private (even commercial) interests and bypassing nation states and their democratic institutions. Other commentators regard MSC's criteria as not strict enough (as applied) (Constance and Bonanno, 2000). Another comment concerns the fact that MSC tends to appeal more to Northern retailers (Bush, 2010) and to middle- and upper-class consumers (Iles, 2004) and much less to poorer producers and consumers. Its chosen strategy may have distributional side effects, as the (high) costs involved in labelling and certification may marginalize poorer and smaller fishers (Vandergeest, 2007; Boström and Klintman,

2008).[10] Involvement of artisanal fishers is even more complicated, as the unit of certification is the fishery in its entirety and not the individual fisher, so fishers sometimes cannot obtain certification because they compete in the same fishery with large-scale fishing units that refuse to cooperate. Other points of criticism include the lack of inclusion of social issues, such as labour conditions and consequences for the wider local community, in the certification procedures.

Nevertheless, MSC has established an innovative network involving the different social actors engaged in capturing and trading (the products of) a particular fish stock and thereby reduced the negative environmental impacts of uncontrolled globalized fish production and consumption. These actors are operating at different geographical levels of scale, but their common interests converge around sustainable fisheries. Through the MSC label, a flexible arrangement is introduced that seeks to combine environmental protection at the local spaces of production (the fishery practices), while acknowledging the existence of global markets for fish operating through the space of flows. This way, a new, 'green' product identity is created in the otherwise-anonymous global fish market.

### *Other fish labels*

Next to MSC, several other fish labels have recently been introduced to increase the sustainability of particular fisheries, such as FOS, KRAV, Marine Eco-Label Japan (MELJ), Wildfish and Fair-Fish. Each has its own particular features, but most follow the FAO Guidelines for Eco-labelling (Accenture, 2009; Parkes et al, 2009). Except for FOS, they have not yet captured a significant portion of the market.

The FOS certification scheme certifies, with the same seal of approval, both farmed and wild-caught products. FOS is becoming a significant sustainable seafood certification scheme, competing with MSC, especially in Southern and small-scale fisheries. The Swedish environmental labelling organization, KRAV, developed an alternative fisheries labelling scheme out of dissatisfaction with the MSC label because it pays insufficient attention to the feasibility of certification for fishers. KRAV's standards give more specific attention to the sustainability of fishing boats and their gear, compared with MSC (Thrane et al, 2009). MELJ recognizes the role of Japan as one of the largest markets for fishery products and the need for Japanese stakeholders to respond proactively and establish their own eco-labelling scheme.[11] MELJ is mainly a verification scheme to ensure that management systems are in place, but it has not been used very widely. Naturland addresses in its Wildfish eco-label standards not only the responsible management of natural resources and the protection of the entire aquatic ecosystem, but also the social aspects of fishery. An interesting scheme that aims to take social and economic interests in fisheries certification one step further is the Fair-Fish label (see Box 8.4).

## Box 8.4 Fair-Fish

Founded in Switzerland in 2000 by animal-welfare organizations, Fair-Fish differs fundamentally from certification schemes that focus only on sustainability. Fair-Fish includes environmental, social and animal welfare criteria, offering better opportunities to small-scale fisheries, which usually cannot access the world market. Fair-Fish includes animal welfare criteria for fisheries:

> *We accept ... fishing methods which do not hold the fish for a long time in the fishing gear and which allow to stun and kill every fish immediately after it is taken off the water. Traditional fishing at coasts and on lakes can cope with these criteria with good will and suitable methods. Industrial fishing however will hardly be able to keep up.*[12]

The certification sets fair trade conditions and guarantees a fixed minimum price for fishers and their families, while at the same time seeking to conserve fish stocks. Since 2004 the initiative has been involved in a project in Senegal to export 'fair-fish' from coastal fisher-folk to Europe. The first small imports from Senegal began in March 2006 for direct marketing through Migros supermarkets in Switzerland.

*Source:* Accenture (2009)

### *Labelling through aquaculture dialogues*

Most fish labels until now have been restricted to capture fisheries, but increased production from aquaculture gave rise to sustainability concerns leading to an initiative to certify and label this activity as well. WWF has set up forums to develop sets of principles for responsible aquaculture production that will subsequently form the basis of farm certification standards. For eight different aquaculture species (e.g. shrimp, salmon, tilapia, pangasius), a group of key stakeholders, including producers, buyers and retailers, NGOs and scientists, is engaged to set standards of best practices (Belton et al, 2010) (see Box 8.5 for an example related to tilapia).

It is still too early to conclude whether the various aquaculture dialogues will have the expected impacts and result in reliable certification schemes. Nevertheless, these dialogues may be the necessary first step towards increasing the sustainability of aquaculture production. Important challenges remain, however, on how to involve smallholder producers, how to make comparisons across different specific local contexts and, especially, how to engage the people in Asia, as they consume the larger portion of the fish produced but thus far seem little interested in its sustainability.

### *Discussing fish labelling*

Voluntary, market-based labelling of fish is expanding fast and has become established as an important tool in global environmental fisheries governance. This growth has also led to debates on their effectiveness in promoting sustainable fish stocks, their legitimacy and how to deal with the presence of multiple competing labels.

### Box 8.5 Tilapia farm in Honduras complying with the International Standards for Responsible Tilapia Aquaculture

The Regal Springs Tilapia farm Aquafinca in Honduras was the first aquaculture operation to be audited by the Institute for Marketecology (IMO) for compliance with International Standards for Responsible Tilapia Aquaculture (ISRTA), which were completed in December 2009 by the Tilapia Aquaculture Dialogue. This was the first of a number of 'pilot' audits scheduled under an agreement between WWF and GlobalGAP. The audit was performed in 2010 by auditors from IMO, a worldwide operating certification organization based in Switzerland that has been accredited for auditing against the GlobalGAP aquaculture standards and complies with requirements for Certification Bodies auditing against the ISRTA. When the Aquacultural Stewardship Council (ASC) launches its consumer label in 2011, this farm will be among the first to receive this ASC label.

*Source:* ASC (2010)

Whether eco-labelling really leads to more sustainably managed fish stocks, depends, at least in part, on the professionalism of NGO staff and certifying bodies in determining the relevant sustainability criteria and applying them correctly (Belton et al, 2010). Their level of professionalism is particularly relevant when dealing with complex fisheries, where knowledge gaps may exist and different interests must be combined. In this way, fish labelling organizations are faced with a dilemma because, on the one hand, standards and their application should be rational, robust and capable of withstanding close scrutiny from competitors, consumers and other interested parties; however, on the other hand, if standards are strictly applied, they allow for the certification of only the already more sustainably managed fisheries and not endangered ones, hence limiting their contribution to increasing sustainable fisheries. In addition, strict application of standards requires the availability of sufficient and reliable information, and this tends to exclude data-poor fisheries, which are mainly found in developing countries. In addition, some commentators argue that labels ultimately will not be able to solve the problem of sustainability in fish provision because they do not address the fundamental issue of over-consumption.

Contrary to official regulations, the legitimacy of fish labels is based less on formal democratic procedures, and more on their (intended) output. Certification schemes claim that their authority derives from the use of objective, scientific methods, applied in a transparent manner to achieve the publicly recognized goal of assuring the future sustainability of fish provision with the participation of all relevant stakeholders. Some argue that such labels are undermining the responsibility and authority of governments in fisheries' governance. Others point to the risk of using scientific information selectively and the potential bias towards private interests in the certification process because both NGOs and certifying agents have an interest in increasing the number of certified fisheries.

Furthermore, commentators criticize the presence of multiple private labelling schemes, as illustrated above, because this may lead to confusion among consumers as well as producers. Each scheme certifies slightly different elements, applies different standards and uses different methodologies. Hence, consumers may face difficulties when choosing a labelled fish product because they cannot be completely sure what each label actually stands for (Iles, 2007). As consumers may have different sustainability concerns, standardizing labelling schemes is complicated. Producers also find it problematic to select a particular label. The fishing industry generally bears the costs of certification and meeting any conditions imposed during the process. This means it will select only one scheme based on an assessment of the potential costs and benefits involved, combined with how well it is recognized on the market. As a result, the producers' choice needs not necessarily be the scheme that is best performing in terms of environmental concerns.

Finally, certification schemes may have unintended social and economic consequences. As mentioned before, small-scale and data-poor fisheries may not be able to provide the required data and are consequently denied access to attractive markets for selling their harvests. In addition, poor consumers may have trouble accessing the more expensive sustainably labelled fish and hence have to rely on non-labelled fish products, which may have more negative consequences for the environment.

## Supply-chain actors engaged in promoting sustainability

Several fish processing, food service and retailing companies have also committed themselves to promote more sustainable fisheries (Parkes et al, 2009; UNEP, 2009). For instance, Unilever, a large food-processing company, was already involved in launching MSC and later committed to source all its fish from sustainably managed fisheries. The company already asks its suppliers to confirm that their fish are legally captured in specified areas and that they are not involved in catching species threatened with extinction.

A substantial number of large supermarket chains, such as Tesco, Walmart, Carrefour, Ahold and Kaufland, have declared their commitment to support sustainable fisheries, often by selling MSC-certified products (see Box 8.6 for the examples of Sainsbury's and Carrefour). In December 2007, the Dutch retail sector as a whole agreed to work towards selling only sustainable fish and seafood in its more than 4500 stores by 2011.[13]

## Fish wallet-cards: Involving consumers in global environmental governance

Environmental NGOs have introduced another instrument to promote sustainable fisheries. Through consumer guides, or 'fish wallet-cards', people are encouraged to engage in protecting marine life by making environmentally sound choices when shopping for seafood (Oosterveer and Spaargaren, 2011). As the Monterey Bay Aquarium explained: 'Fishing practices worldwide are

## Box 8.6 Sainsbury's and Carrefour's sustainable fish procurement policies

Sainsbury's (UK) developed a decision tree together with stakeholders to decide what fisheries they will source from. They chose to make the 'big 5' (cod, haddock, salmon, tuna and prawns) the focus of their efforts, as they account for 80 per cent of all fish sold every week, and to move them to 100 per cent sustainable sourcing, ensuring that the fish are caught or reared with minimal impact on stocks, ecosystems and the wider environment. Sainsbury's also has projects working with suppliers, fishers, vessel owners and governments in Sri Lanka and the Maldives to make fresh and canned tuna fully MSC certified. The company requires that 100 per cent of its canned tuna comes from pole-and-line capturing methods.

The French Carrefour Group initiated its own responsible fishing standard in 2004, 'pêche responsable'. The standard was applied only to selected species fished in Iceland and sold in Carrefour supermarkets in France and Belgium. It assured responsible fishing, and it guaranteed optimal traceability and stock management, as well as respect for the ecosystem. Despite these efforts, however, this initiative never became very successful, and today Carrefour is progressively withdrawing its own standard and promoting MSC and FOS-certified items instead.

*Source:* Accenture (2009); Parkes et al (2009)

damaging our oceans – depleting fish populations, destroying habitats and polluting the water. Informed consumers can help turn the tide.'[14] And, the Seafood Choices Alliance strives to:

> *mobilize market forces in the global seafood sector, catalyzing positive action in support of ocean conservation. Our vision is of a global market where all stakeholders share responsibility, with minimal negative impact on the oceans, or on the ecosystems or communities that depend on them.*[15]

Although these wallet-cards come in different forms, they all differentiate between species that can be bought without special environmental concerns and those that should be avoided, using a traffic light system. The consumer-oriented information on some of these wallet-cards is still rather incomplete and not well connected to information flows from supply-chain procedures and actors, but interested consumers can access a website for more detailed information. For each species, the website features data about its main sources, methods of capture and its environmental condition, sometimes supplemented with recipes and nutritional guidance.

With these consumer wallet-cards, NGOs have taken every effort to develop a tool that consumers can easily read, using a simple distinction between good and bad, and referring only to fish available on the local market. Like the MSC label, the fish wallet-card is an attempt to reconnect the provision of fish with its practices of consumption, this time starting at the consumer end of the chain. Here again, shifts in consumer preferences are expected to lead to adaptations in both global practices of food retail and in local practices of capturing marine fish. Nevertheless, fish wallet-cards are confronted with the same comment: that it is a useful tool only for consumers that have a choice when buying fish, which limits their effectiveness as an instrument to promote sustainable fisheries.

## Conclusions

Fish production and consumption are increasingly globally organized, so nation state-based regulations are inadequate to combine fish provision for a growing global demand with the need to secure the future sustainability of fisheries' resources. Innovative governance arrangements that more effectively realize this aim have become a necessary complement. After presenting the important challenges in securing the sustainability of global fish provision, this chapter has illustrated several attempts to fill this regulatory gap.

The new arrangements introduced are based on multi-level and multi-actor involvement in global environmental governance. Fish labels and fish wallet-cards make extensive use of market-based logics and imply a view that environmental NGOs and consumers are co-regulators and drivers for change. These instruments try to make adequate and reliable information about the production, processing and trading practices involved in fish provision available to consumers concerned about its sustainability. These environmental NGOs, operating at the global level, are able to cross-cut national borders and cultures and sometimes even trespass on individual nation states when trying to implement parts of their global sustainability agenda. They seem best capable of creating trust among different groups of actors involved in fish supply chains, which is complicated because they extend over long distances in time and space. In doing so, environmental NGOs are not just complementing but sometimes even replacing nation state agencies and private companies in their regulatory role.

Although the introduced private-governance arrangements may be weak in their formal democratic legitimacy, they could play an important role in preventing and reducing environmental deterioration, provided they interact with governmental arrangements. The instruments of government-sanctioned marine reserves, rules-based restrictions on access to fish resources, stringent distributive schemes and the curtailment of illegal, unregulated and unreported fishing through governments remain essential. In this respect, international agreements should aim at protecting fishery resources rather than promoting fish trade. Multilateral coordination in this respect is required to secure the availability of fish as a source of healthy food for future generations, but these efforts should be harmonized with private initiatives.

*Take-home lessons*

- The global fish supply illustrates the need to incorporate sustainability into trade practices and not to rely only on regulating producer practices.
- Multilateral institutions such as the WTO may provide a platform for addressing sustainability challenges at the global level, but its institutional framework often makes it difficult to do so effectively.
- The absence of clear standards for private certification may lead to the introduction of multiple schemes that compete on sustainability. This is not necessarily negative but may confuse producers and consumers.

## Notes

1 This is compared with less than about 10 per cent of global meat output entering international trade.

2 Under very exceptional conditions, however, activities involved in capturing and processing fish may be included, as illustrated in the case of the shrimp–turtle dispute (FoEI, 2001).

3 Nevertheless, private fish quality and safety schemes constitute an increasing challenge for international fish trade. Developed countries' importers impose high and complex quality requirements, as well as health and safety requirements, on producers (Ponte et al, 2007). Developing countries fear that such requirements are yet another barrier for entry into the lucrative global fish markets (Potts and Haward, 2007).

4 See the WTO Ministerial Declaration at the Hong Kong meeting, 18 December 2005, Annex D, article 9.

5 The eco-label index (www.ecolabelindex.com/ecolabels, accessed 10 August 2010) counts 328 different eco-labels, of which 12 are categorized under fish/fisheries.

6 For instance, ISO, the largest standard-developing organization, also intends to promote the sustainable development of the fisheries and aquaculture sectors (Accenture, 2009).

7 See www.msc.org/about-us/vision-mission (accessed 16 February 2011).

8 According to the MSC (www.msc.org/track-a-fishery/what-is-a-fishery, accessed 16 February 2011), a 'fishery' may include one or more 'units of certification'. A unit of certification is usually defined by reference to the following:
- target fish species and stock;
- geographic area of fishing;
- fishing method, gear, practice and/or vessel type.

9 See www.msc.org (accessed 22 December 2010).

10 In particular, the small-scale sector of developing countries has difficulties in providing the environmental data that constitute the basis for a certification process and may lack sufficient capacity to comply with the technical or financial requirements

involved. These unintended consequences are the reason why MSC is introducing a special certification procedure for 'data-poor' fisheries (MSC, 2009).

11 Japan knows many small-scale fishers and fishing boats, as well as a wide variety of target fish species in fisheries. An official framework encourages fishers and others users of the resources to manage them (Accenture, 2009).

12 See www.fair-fish.ch/files/pdf/english/ff_short.pdf (accessed 13 August 2010).

13 Until then, fish products are assigned a sticker informing the consumer that the retail sector is making steps to realize a full assortment of sustainable fish products in the near future.

14 See www.montereybayaquarium.org/cr/seafoodwatch.asp (accessed 22 December 2009).

15 See www.seafoodchoices.com/whoweare (accessed 23 December, 2009).

## Further Reading

Kurlansky, M. (1999) *Cod: A Biography of the Fish that Changed the World*, Vintage Books, London: a classic book on the collapse of the cod fisheries near Newfoundland.

Myers, R. and Worm, B. (2003) 'Rapid worldwide depletion of predatory fish communities', *Nature*, vol 423, pp280–283: offers concrete information on the degradation of the worlds' fisheries.

Naylor, R., Goldburg, R., Primavera, J., Kautsky, N., Beveridge, M., Clay, J., Folke, C., Lubchenko, J., Mooney, H. and Torrellet, M. (2000) 'Effect of aquaculture on world fish supplies', *Nature*, vol 405, pp1017–1024: an overview of the environmental impacts from aquaculture.

## References

Accenture (2009) *Assessment of On-pack, Wild-capture Seafood Sustainability Certification Programmes and Seafood Labels*, Accenture and WWF International, Zurich

Ahmed, M. (2006) *Market Access and Trade Liberalisation in Fisheries*, ICTSD, Geneva

Allison, E. (2001) 'Big laws, small catches: Global ocean governance and the fisheries crisis', *Journal of International Development*, vol 13, pp933–950

ASC (Aquaculture Stewardship Council) (2010) 'Press statement', 22 October, www.ascworldwide.org

Belton, B., Murray, F., Young, J., Telfer, T. and Little, D. (2010) 'Passing the Panda standard: A TAD off the mark?', *AMBIO*, vol 39, no 1, pp2–13

Boström, M. and Klintman, M. (2008) *Eco-Standards, Product Labelling and Green Consumerism*, Palgrave MacMillan, Houndmills

*Bridges Weekly* (2007) 'Members clash over how to discipline fisheries subsidies', *Bridges Weekly*, vol 11, no 22, ICTSD, Geneva

*Bridges Weekly* (2008) 'Rough sailing for fisheries subsidy talks', *Bridges Weekly*, vol 12, no 4, ICTSD, Geneva

Bush, S. (2010) 'Governing "spaces of interaction" for sustainable fisheries', *Tijdschrift voor economische en sociale geografie*, vol 101, no 3, pp305–319

Bush, S. and Minh, L. (2005) *Fish Trade, Food and Income Security: An Overview of the Constraints and Barriers Faced by Small-scale Fishers, Farmers and Traders in the Lower Mekong Basin*, Oxfam America, Phnom Penh

Bush, S. and Oosterveer, P. (2007) 'The missing link: Intersecting governance and trade in the space of place and the space of flows', *Sociologia Ruralis*, vol 47, no 4, pp384–399

Constance, D. and Bonanno, A. (2000) 'Regulating the global fisheries: The World Wildlife Fund, Unilever and the Marine Stewardship Council', *Agriculture and Human Values*, vol 17, pp171–187

FAO (Food and Agriculture Organization of the United Nations) (2000) *International Trade in Fishery Products and the New Global Trading Environment*, FAO, Rome

FAO (2005) *Guidelines for the Ecolabeling of Fish and Fishery Products from Marine Capture Fisheries*, FAO, Rome

FAO (2006) *The International Fish Trade and World Fisheries*, FAO, Rome

FAO (2009) *The State of World Fisheries and Aquaculture 2008*, FAO, Rome

FoEI (Friends of the Earth International) (2001) *Sale of the Century? The World Trade System: Winners and Losers*, Friends of the Earth International, Amsterdam

Garcia, S., Cochrane, K., Santen, G. and Christy, F. (1999) 'Towards sustainable fisheries: A strategy for FAO and the World Bank', *Ocean & Coastal Management*, vol 42, no 5, pp369–398

Gulbrandsen, L. (2009) 'The emergence and effectiveness of the Marine Stewardship Council', *Marine Policy*, vol 33, no 4, pp654–660

Hoof, L. (2010) *Who Rules the Waves?*, WUR, Wageningen

ICTSD (International Centre for Trade and Sustainable Development) (2007) 'WTO: Members discuss details of US proposal on fisheries subsidies', *Bridges Trade BioRes News, Events and Resources at the Intersection of Trade and Biodiversity*, vol 7, no 9, pp4–6

Iles, A. (2007) 'Making the seafood industry more sustainable: Creating production chain transparency and accountability', *Journal of Cleaner Production*, vol 15, pp577–589

Juda, L. (2002) 'Rio plus ten: The evolution of international marine fisheries governance', *Ocean Development & International Law*, vol 33, pp109–144

Love, P. (2010) *Fisheries: Will Stocks Last?*, OECD, Paris

Mansfield, B. (2006) 'Assessing market-based environmental policy using a case study of North Pacific fisheries', *Global Environmental Change*, vol 16, no 1, pp29–39

MSC (Marine Stewardship Council) (2002) *Fish 4 Thought; The MSC Quarterly Newsletter*, vol 2, MSC, London

MSC (2009) *Risk-Based Framework and Guidance to Certification Bodies (Version 1)*, MSC, London

MSC (2010) *Annual Report 2009/10*, MSC, London

OECD (Organisation of Economic Co-operation and Development) (2010) *Globalisation in Fisheries and Aquaculture: Opportunities and Challenges*, OECD, Paris

Oosterveer, P. and Spaargaren, G. (2011) 'Organising consumer involvement in the greening of global food flows: The role of environmental NGOs in the case of marine fish', *Environmental Politics*, vol 20, no 1, pp97–114

Parkes, G., Walmsley, S., Cambridge, T., Trumble, R. , Clarke, S., Lamberts, D., Souter, D. and White, C. (2009) *Review of Fish Sustainability Information Schemes: Final Report*, MRAG and Stirling University, Edinburgh

Peterson, M. (1993) 'International fisheries management', in P. Haas, R. Keohane and M. Levy (eds) *Institutions for the Earth: Sources of Effective International Environmental Protection*), The MIT Press, Cambridge, MA, pp249–305

Ponte, S., Raakjaer, J. and Campling, L. (2007) 'Swimming upstream: Market access for African fish exports in the context of WTO and EU negotiations and regulation', *Development Policy Review*, vol 25, no 1, pp113–138

Potts, T. and Haward, M. (2007) 'International trade, eco-labelling, and sustainable fisheries: Recent issues, concepts and practices', *Environment, Development and Sustainability*, vol 9, no 1, pp91–106

Roheim, C. and Sutinen, J. (2006) *Trade and Marketplace Measures to Promote Sustainable Fishing Practices*, ICTSD, Geneva

Rubik, F. and Frankl, P. (eds) (2005) *The Future of Eco-labelling: Making Environmental Product Information Systems Effective*, Greenleaf, Sheffield

Sriwichailamphan, T. (2007) *Global Food Chains and Environment: Agro-food Production and Processing in Thailand*, WUR, Wageningen

Tacon, A. (2010) 'Climate change, food security and aquaculture', in OECD (ed) *Advancing the Aquaculture Agenda: Workshop Proceedings*, OECD, Paris, pp109–119

Thrane, M., Ziegler, F. and Sonesson, U. (2009) 'Eco-labelling of wild-caught seafood products', *Journal of Cleaner Production*, vol 17, pp416–423

UNEP (United Nations Environment Programme) (2009) *Certification and Sustainable Fisheries*, UNEP, Paris

Vandergeest, P. (2007) 'Certification and communities: Alternatives for regulating the environmental and social impacts of shrimp farming', *World Development*, vol 35, no 7, pp1152–1171

# Section III

# Future Perspectives

# 9

# Roles of Producers in Sustainable Food Provision

This chapter aims to:

- discuss the roles of food producers in food-supply chains and agricultural policies;
- present different producer strategies through promoting GM crops, regional food, organic farming and food sovereignty;
- review several contributions producers can offer to enhance sustainable food provision in the future.

## Introduction

Farmers nowadays find themselves operating in a globalizing world of food and are faced with multiple challenges, not only to secure the future of their farms, their livelihoods and the local community, but also to contribute to more sustainable food provision. Farmers must respond to these challenges while playing their role as part of large supply chains dominated by corporate traders, processors and retailers. In reacting to such pressures and demands, farmers develop different strategies. To reduce environmental impacts, some focus on using high-tech methods, while others try to embed their production practices in local communities and local eco-systems. For this chapter, we selected some of the producer strategies that figure prominently in scientific and public debates on the future of farming and food, to illustrate some of the issues that involve globalization and sustainability.

We first discuss genetic modification as a recent technological innovation, followed by regional food labels and multi-functionality, and finally we extensively review organic agriculture as a strategy towards increasing sustainability. We consider the perspectives of small-scale farmers when engaging in global trade and introduce food sovereignty as a producer strategy. The chapter concludes with a brief recapture of the main aims that future producer strategies should entail.

## Food Producers, Food Politics and Food-Supply Chains

Farmers are faced with a continual pressure to increase their productivity so as to maintain their position in the supply chain. In the EU, the price guarantees that previously existed under the CAP have gradually disappeared since the 1980s, resulting in increased price volatility and a downward trend in producer prices. For this and other reasons, the number of farmers in Europe has declined considerably. The remaining farmers are becoming increasingly bifurcated, such that one group applies highly intensive production methods and another group combines agriculture with activities such as tourism and direct food sales, as well as environmental, social or healthcare activities.

In the US, similar trends occur, but interestingly, the total number of farms did not decline but increased by 4 per cent during the period from 2002 to 2007. Many of those new farmers were, however, small scale, as 36 per cent of the total number of farms were registered as residential/lifestyle and 21 per cent as retirement, all with sales of less than $250,000 annually. During this period, farmers became more diverse, and more women and members of different ethnic groups became principal farm operators (USDA, 2009).

Nearly all farmers in more developed countries make extensive use of external inputs in farming, such as seeds, chemicals and fertilizers, which they buy from agrochemical and seed companies. Over time, farmers have become increasingly dependent on just a few private companies because the market for agricultural inputs has become concentrated very quickly. In 2007, for example, over 80 per cent of the seeds sold on the global seed market were sold under

**Table 9.1** *Ten largest companies on the global seed market (2007)*

| | *Company* | *Seed sales ($ million)* | *% of global market for branded seed* |
|---|---|---|---|
| 1 | Monsanto (US) | 4964 | 23 |
| 2 | DuPont (US) | 3300 | 15 |
| 3 | Syngenta (Switzerland) | 2018 | 9 |
| 4 | Groupe Limagrain (France) | 1226 | 6 |
| 5 | Land O'Lakes (US) | 917 | 4 |
| 6 | KWS AG (Germany) | 702 | 3 |
| 7 | Bayer Crop Science (Germany) | 524 | 2 |
| 8 | Sakata (Japan) | 396 | <2 |
| 9 | DLF-Trifolium (Denmark) | 391 | <2 |
| 10 | Takii (Japan) | 347 | <2 |
| | TOTAL | 14,785 | 67 |

*Source:* ETC (2008)

**Table 9.2** *Ten largest agrochemical companies (2007)*

| | *Company* | *Agrochemical sales ($ million)* | *% of global market for agrochemical sales* |
|---|---|---|---|
| 1 | Bayer (Germany) | 7458 | 19 |
| 2 | Syngenta (Switzerland) | 7285 | 19 |
| 3 | BASF (Germany) | 4297 | 11 |
| 4 | Dow AgroScience (US) | 3779 | 10 |
| 5 | Monsanto (US) | 3599 | 9 |
| 6 | DuPont (US) | 2369 | 6 |
| 7 | Makhteshim Agan (Israel) | 1895 | 5 |
| 8 | Nufarm (Australia) | 1470 | 4 |
| 9 | Sumitomo Chemical (Japan) | 1209 | 3 |
| 10 | Arysta Lifescience (Japan) | 1035 | 3 |
| | TOTAL | 34,396 | 89 |

*Source:* ETC (2008)

brand names, and nearly 70 per cent of this share was in the hands of just ten companies. Tables 9.1 and 9.2 illustrate the level of concentration and the global spread of companies' origins. In public debates, Monsanto, the largest seed company and the fifth largest agrochemical company, has become a public icon as exemplifying the role of private companies in agriculture.

Despite much criticism regarding the political and economic power concentrated in these companies, the industrialized/modern form of developing food provision, which is based on applying more external inputs, has contributed to hugely increased yields over the last 60 years. Green et al (2003, p148), however, argue that 'there is no single logic that drives the development of food systems and different societies exhibit a complex and contingent mix of different food supply systems at different levels of definition and development'. Next to organic food, they identify a 'new industrial' strategy as an alternative to agro-industrialization. This new industrial paradigm pleads for using modern technology in intensive methods of crop management (including the application of biotechnology, information and communication technology and the use of new materials). This strategy is oriented towards high-production output and low-labour input but applies eco-sensitive approaches to biodiversity, with an emphasis on food hygiene and quality. So, simply opposing large-scale industrial corporate agriculture and small-scale organic farming does not do justice to the diversity in strategies aimed at increasing sustainability.

## Farmers and technological innovation: The case of GMOs

A recent technological innovation, genetic modification leading to GMOs and crops, has generated widespread public debates. GM crops are the result of scientific innovations in the 1970s and 1980s, leading to the first outdoor tests in 1986. In 1994, the first GM crop was available on the market, the FlavrSavr tomato, which had a longer shelf-life than conventional tomatoes. After this, many more GM crops were developed, mostly with the intention to improve

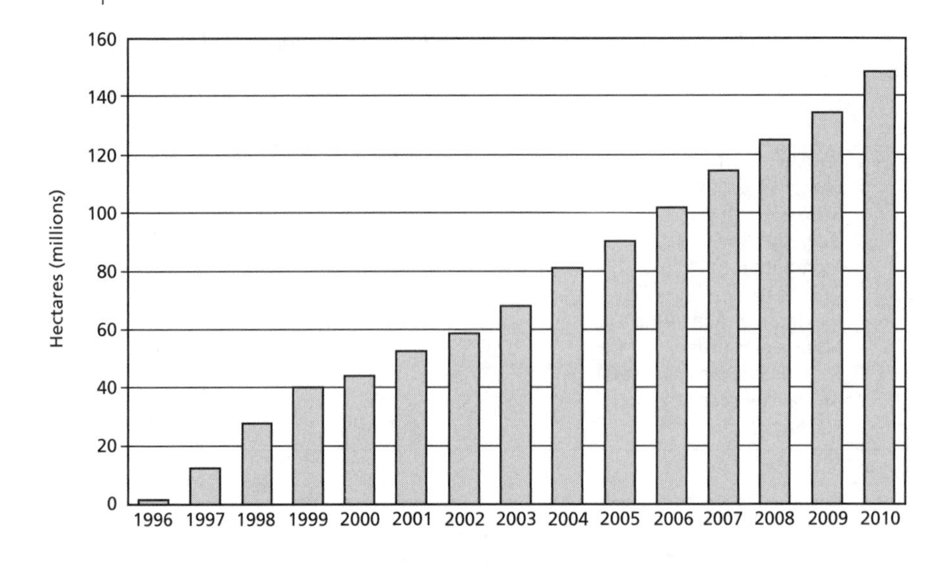


**Figure 9.1** *GM crops worldwide (million hectares)*

*Source:* ISAAA (2010)

agronomic traits, such as reduced use of chemicals, by building in resistance against pesticides, herbicides or fungicides in the crops. Other traits were also inserted into the genetic composition of crops, for example, to enhance the production of vitamins and other useful substances that are naturally present.

Although the FlavrSavr tomato is no longer cultivated, many other GM crops are being produced, and in the 15 years since its first introduction, the area for GM crops has expanded very rapidly to 134 million hectares in 2009, spread over 25 countries and involving 14 million farmers. Figure 9.1 depicts this rapid growth.

Despite the expansion of the global area used for planting GM crops, the US is still dominating the cultivation of these crops. As Table 9.3 shows, the country is still responsible for nearly half of the total area, although producers in Brazil, Argentina and India have substantially increased their use of GM technology as well.

European countries do not figure in Table 9.3, but although some may expect these countries not to produce GM crops at all, some countries do, although in smaller areas compared with the larger producers (see Table 9.4 for details).

The striking difference between the US and the EU when producing GM crops is mostly explained by the much smaller scale of agriculture in Europe but primarily by the higher sensitivity for food safety issues among the continent's consumers. This has also led to different GM-food policies (see Box 9.1).

The extent of GM technology ranges widely among various crops. Although GM has come to dominate soybean production worldwide where now nearly 80 per cent of the area is planted with GM crops (see Table 9.5), there is no application for the most important food crops: rice and wheat. For cotton, GM

Table 9.3 *Global area of GM crops, 1997–2009*

| *Country* | *1997* | | *2000* | | *2009* | |
|---|---|---|---|---|---|---|
| | *Area (million hectares)* | *% global area* | *Area (million hectares)* | *% global area* | *Area (million hectares)* | *% global area* |
| US | 8.1 | 64 | 30.3 | 69 | 64.0 | 47.8 |
| Brazil | – | | – | | 21.4 | 15.9 |
| Argentina | 1.4 | 11 | 10.0 | 7 | 21.3 | 15.9 |
| India | – | | – | | 8.4 | 6.2 |
| Canada | 1.3 | 10 | 3.0 | 7 | 8.2 | 6.1 |
| China | 1.8 | 14 | 0.5 | 1 | 3.7 | 2.8 |
| Paraguay | – | | – | | 2.2 | 1.6 |
| **Total*** | **12.8** | **100** | **44.2** | **100** | **134** | **100** |

*Note:* * Includes production in countries not specified in the table.

*Source:* ISAAA (2009)

Table 9.4 *European area of GM crops (maize unless indicated otherwise )(hectares)*

| *Country* | *Area 2006* | *Area 2008* | *Area 2009* |
|---|---|---|---|
| Spain | 55,667 | 79,269 | 76,057 |
| Czech Republic | 1290 | 8380 | 6480 |
| Romania | 90,000[i] | 7146 | 3994 |
| Portugal | 1250 | 4851 | 5202 |
| Germany | 947 | 3137 | 0 |
| Poland | 100 | 3000 | 3000 |
| Slovakia | 30 | 1900 | 875 |
| **Total** | **149,284** | **107,683** | **95,608** |

*Note:* [i] Romania had a substantial area of GM soybean production, but this had to be given up when the country joined the EU in 2007 because this was not allowed within the European Community.

*Source:* GMO Compass (2008); FoEI (2010)

technology is applied in nearly 50 per cent of the total planted area, and for maize and canola, these shares are about 20 to 25 per cent. Interestingly, these crops have been dominant in this development since the introduction of GM technology.

Over the last two decades, the introduction of GM crops has been a very controversial subject. Some consider GM technology an attractive innovation because farmers will benefit from higher yields and lower costs through reduced

Table 9.5 *Worldwide cultivation of GM crops (2009)*

| *Crop* | *Area (million hectares)* | *Area GM (million hectares)* | *% GM of total area* |
|---|---|---|---|
| Soybean | 90 | 69.3 | 77 |
| Maize | 158 | 41.1 | 26 |
| Cotton | 33 | 16.2 | 49 |
| Canola/rapeseed | 31 | 6.5 | 21 |

*Source:* ISAAA (2009)

### Box 9.1 The dispute between the US and the EU on GM food

The US considers GM crops not to be fundamentally different from conventional crops (they are called '*substantially equivalent*'). Therefore, there is no need for specific measures for the category as a whole, but only for those elements that are changed. When such modifications test as safe, the products do not need to be segregated or labelled. The EU, in contrast, views GM crops as different from conventional crops. Hence, the EU requires labelling of food containing more than 0.9 per cent GMOs, thereby obliging exporters trading such products to the EU to also adhere to this requirement.

These differences in GM-food regulation have created problems in international trade. The US considered the EU-labelling requirement to be unacceptable under the WTO because it created a non-tariff trade barrier with potentially large trade-distorting effects, while being imposed 'without sound scientific evidence'. Based on these considerations, the US (together with Canada and Argentina) filed a complaint against the EU within the WTO. The WTO Dispute Settlement Panel concluded that the EU policy (and its discussions) resulted in unacceptable delays in approval procedures and that national bans (by France, Germany, Austria, etc.) were WTO-non-compliant. The panel expressed no judgement on the 'lack of scientific evidence' for the EU's policy, however, and thus the labelling requirement itself was not rejected. The matter was resolved when Canada (2009) and Argentina (2010) agreed to have regular, bilateral meetings with the EU to discuss GM agriculture. The EU and US aim to come to a similar accord.

*Source: Bridges Weekly*, ICTSD, Geneva (various issues)

pesticide spraying, while their flexibility in labour usage increases. Further innovations could even improve the productivity of crops under harsh conditions, such as extensive drought or saline soils. Environmental benefits are also expected through reduced pesticide use and alleviating pressure on the remaining natural areas. Finally, some also expect a reduction of hunger in the world, as more food would be available on a secure basis. The quality of food could also be higher and better adapted to consumer and retailer demands.

Others, environmental social movement advocates in particular, stress the risks associated with using GM technologies. They point to potential human health, environmental and socio-economic consequences resulting from these technologies. Unexpected toxins and unknown allergenic proteins may be present in food, and unintended genes spread to other crops, threatening biodiversity. GM technology may increase farmers' dependency on large agrochemical companies even further, as they need to buy the seeds in combination with specific pesticides.[1]

Until now, farmers in developing countries hardly applied GM technology, except in the production of cotton. Nevertheless, some proponents (e.g. Borlaug, 2000) expect GM crops to solve the hunger problem by increasing productivity and solving specific nutrition problems (see Box 9.2 for the example of golden rice).

Researchers have concluded that although GM crops may generate benefits for producers, consumers and the environment, it will take many more years (10 to 15) than the three to five years initially expected (Eicher et al, 2006). Raney

### Box 9.2 Golden rice

In 2000, the Swiss biotech company Syngenta discovered 'golden rice', a GM rice strain that contains more beta-carotene than conventional rice. Beta-carotene is converted into vitamin A in the human body and thereby prevents blindness. Golden rice would be offered for free to farmers in Southeast Asia where (vitamin A-deficient) blindness is common. Since its discovery, golden rice has been used as a showcase to illustrate the potential of GM technology for developing countries, yet it has not been introduced in the farmers' rice fields so far.

In 2002, Dawe (2002) observed several unresolved problems, which seem to remain unresolved. First, the quantity of vitamin A available in golden rice should be increased to be effective, especially if people do not completely switch to golden rice because of dietary preferences. Second, attaining widespread consumer acceptance of this very different-looking commodity is essential, but consumers have been ignored in this process. Third, it is uncertain whether beta-carotene will still be available after longer-term storage and preparation. Finally, if the additional costs of producing golden rice result in higher prices, its attractiveness as a vehicle for decreasing vitamin A deficiency will be reduced.

*Sources:* www.gmo-compass.org/features (accessed 17 December 2007); www.foodnavigator.com, 'New golden rice has 20 times more beta-carotene' (accessed 29 March 2005); Dawe (2002)

(2006) and Raybould and Quemada (2010) conclude that farmers in developing countries could also profit from applying GM crops, but this would require a fairly high institutional capacity to ensure that farmers have access to suitable innovations on competitive terms. This makes it unlikely that poor farmers, who are generally the most food insecure, will profit from GM technology.

## Disengaging from Industrial Foods

Using local or regional identity is another strategy to enhance sustainability. Applying local identities strengthens traditional production methods and artisanal processing techniques. Regional products stand out as special in an otherwise anonymous global market with industrial foods. Geographical indications connect food products to specific places of production (each with its own natural conditions, cultural and gastronomic traditions, etc.) and production process (artisanal, traditional, farm-based, etc.) (Renting et al, 2003). Labels of origin enable consumers to appreciate particular quality characteristics and obtain information about the production and processing methods used, which they rarely get from conventional products.

In 1919, France was the first country to establish a national system for protecting regional products through its 'appellation d'origine' label, whereby food that is identified by a specific geographical zone cannot legally be produced elsewhere. Such food is linked to regional specialties with historic roots and includes such characteristics as good taste, diversity, the importance of personal relationships between producers and consumers and a lack of artificial additives or preservatives. Emphasis is placed on the connectivity of particular foods to their specific territory, collective history and savoir-faire in their production and consumption: 'produits de terroir' (DeSoucey and Téchoueyres, 2009).

Over time, the EU has created an encompassing system that allows the registration and legal protection of 'protected designations of origin' (PDOs) and 'protected geographical indications' (PGI) for food products. In recent years, many food products have been registered within the EU, followed by other countries. This trend falls into the broader 'quality turn' that is occurring in the world of food, whereby producers seek to distinguish quality food from standard food. Local producers emphasize the origin and quality of their products to appeal to consumers in distant markets (Friedmann and McNair, 2008) and to sell them for higher prices than their industrially produced competitors. Moreover, local foods help maintain local traditions and cultures because they are collectively owned and may bypass conventional supply chains.

Promoting local food fits in the broader aim to promote multi-functional agriculture, a rather prominent strategy in Europe. Although Sachs and Santarius (2007) argue that multi-functional agriculture is not new because agriculture has always been a polyvalent activity and should not be reduced to the language of crop yields and money, this stance has not been very visible in agricultural politics. Multi-functional agriculture underlines that farming produces not only food, but also livelihoods, landscapes and meaning. Different farming systems in various social contexts have direct impacts on water cycles, soil regeneration, condition the patrimony of plants and animals, and mould hills and valleys. To secure these non-economic contributions from agriculture, multi-functionality must be included in agricultural policies. Markets are unrivalled in producing and delivering goods and services efficiently, but on their own they cannot create community or equity, security or sustainability, sacredness or beauty. It is up to citizens, governments and politicians to ensure that such common goods are carefully safeguarded and reproduced.

## Organic Agriculture

Organic agriculture is another strategy to increase sustainability in food provision while creating a separate market for these products. Organics claim to be a radical break with the conventional, industrial food chain and have grown rapidly in recent decades. The global area for organic farming increased from 11 million hectares in 1999 to more than 35 million hectares in 2008 (see Table 9.6). This area represents 0.82 per cent of all agricultural lands worldwide (Willer and Kilcher, 2010).

Organic farming emerged as an offshoot from Rudolf Steiner's philosophy on biodynamics[2] in Germany of the 1920s, but evolved into a less esoteric approach to farming in the 1960s when taken up by the student and youth protest movements. During the following decades, these efforts became professionalized and an alternative method of producing food emerged. In the early 1980s, organic produce started to appear in supermarkets in North America, Western Europe and Japan. When this was followed by other countries, organics became part of global food-supply chains involving producers and retailers in many countries. The growth of the organic sector was stimulated by food scares and growing

**Table 9.6** *Organic agricultural land, per continent (2008)*

| *Continent* | *Area million hectares* |
|---|---|
| Northern America | 2.5 |
| Latin America | 8.1 |
| Europe | 8.2 |
| Africa | 0.9 |
| Asia | 3.3 |
| Oceania | 12.1 |
| **TOTAL** | **35.1** |

*Source:* Willer and Kilcher (2010)

environmental and trade concerns (Morgan and Murdoch, 2000).

At first, there was no formal definition of organic agriculture, because it was essentially based on general aims and principles that every farmer could apply in their own manner. Over time, however, a complex system of product differentiation developed to distinguish organic from non-organic products. This differentiation was based on third-party certification, whereby an independent, non-interested party verifies whether farming practices conform to organic standards. Certification allowed the organic-food sector to expand and be included in (international) trade and in the range of products offered in supermarkets. The model of certifying organics is unique because unlike most other certification schemes, governments (driven by the US and the EU) set the formal standards (Albersmeier et al, 2009).

According to its global umbrella organization the International Federation of Organic Agriculture Movements (IFOAM), founded in 1972 in Germany, organic agriculture can be defined as:

> *a production system that sustains the health of soils, ecosystems and people. It relies on ecological processes, biodiversity and cycles adapted to local conditions, rather than the use of inputs with adverse effects. It combines tradition, innovation and science to benefit the shared environment and promote fair relationships and a good quality of life for all involved.*[3]

This definition translates into four key principles:[4]

- Health – organic agriculture should sustain and enhance the health of soil, plant, animal, human and planet as one and indivisible.
- Ecology – organic agriculture should be based on living ecological systems and cycles, work with them, emulate them and help sustain them.
- Fairness – organic agriculture should build on relationships that ensure fairness with regard to the common environment and life opportunities.
- Care – organic agriculture should be managed in a precautionary and responsible manner to protect the health and well-being of current and future generations and the environment.

### Box 9.3 Group certification to reduce costs

To reduce the expenses of certification that are unaffordable for small and marginal farmers, the Indian Organic Farming Federation introduced group certification. This Participatory Guarantee System allows certificates to be given to groups that carry out the process of continuous supervision effectively. There are no additional costs involved for getting a certificate, apart from the ones incurred in creating and maintaining the infrastructure for organic farming. All farmers will get a group identity number and a sub-identity number, to keep the system transparent and reliable.

*Source: Indian Express* (2010)

As the market for organic products in Europe expands, its international trade is also increasing, particularly for items that are not (permanently) available for climatic or seasonal reasons. This opens up opportunities for producers in developing countries to export to Europe and the US, provided they are certified against official EU or US standards.[5] Although such a complex organic certification process may be problematic because of the direct and indirect costs involved, they are sometimes supported by special certification schemes developed for groups of farmers (see Box 9.3). Still, the challenges and costs involved mean that larger farmers can profit more easily from these export opportunities.

Exporting organic food requires farmers to develop and maintain high product quality to justify the high prices consumers are asked to pay (Bray et al, 2002). So, there are clear indications that the introduction of organic agriculture technologies enhances farmers' skills in managing resources and maximizing their labour productivity. These improvements in agricultural performance are used in the debate on the potential of organic farming (see Box 9.4).

Organic agriculture development has not been uniform and ranges considerably in various parts of the world. Therefore, it would be useful to describe the dynamics in the US, the EU and Africa in more detail.

## Organic farming in the US

In the US, organic agriculture as a social movement has opposed industrial agriculture and tried to rebuild farming on pre-industrial principles, knowledge and technologies, since its earliest promoters, Henry Wallace and J. I. Rodale. The organic farming movement broadened, especially in the 1960s and 1970s, when small-scale organic agriculture was discovered as a concrete alternative to the chemical-based farming considered responsible for environmental problems, low-quality food and inhumane treatment of animals. Nowadays, organic agriculture has greatly expanded, and many present-day organic farmers are conventional farmers who converted to organic, while others are not completely organic but have mixed operations.

The pioneers initially developed their own understanding of organic farming and communicated this directly to the consumers buying their produce. Over several decades, these individual definitions have evolved through debates,

## Box 9.4 Can organic agriculture feed the world?

Even though only a small part of global food production is presently organic, its potential for reducing global hunger has attracted much attention. Critics argue that by renouncing the use of genetic engineering, chemical fertilizers and pesticides, organic farming will by definition not be able to produce sufficient food to feed the world. They argue that such external inputs are indispensible because otherwise productivity will decrease. Organic-farming supporters respond that technology-intensive agriculture may have increased global food production until today, but at serious environmental and social costs, and thus there is a need for long-term, more sustainable solutions. McIntyre et al (2009) argue that the degradation of ecosystems resulting from conventional intensification strategies now limits or even reverses productivity gains. It is therefore necessary to integrate natural resource management more closely with food production and achieve increased productivity in a sustainable manner, whereby organic principles may provide a valuable framework. Badgley et al (2007) calculated the global potential for organic farming on the basis of data on its recent performance in practice and concluded optimistically that 'organic methods of food production can contribute substantially to feeding the current and future human population on the current agricultural land base, while maintaining soil fertility' (p94).

*Source:* Borlaug (2000); Badgley et al (2007); McIntyre et al (2009)

research and practical experiences, and organic food producers have developed a complicated set of conventions to determine the truthfulness and value of their organic claim. In the 1990s, however, pressure grew to replace the multitude of state and regional standards and develop nationwide ones to gain access to wider markets and prevent misuse if conventional products were sold as organic to profit from the price premium. Hence, in 2002, a national organic standard was implemented.

Despite the advantage of a nationwide standard to reduce confusion among the general public, many accuse this official US Department of Agriculture (USDA) definition of organics of diluting the organic principles. 'A national organic standard was established at the lowest common denominator set of requirements and eliminated competition for the most stringent level of certification' (Howard, 2009, p14). Critics expect this to lead to the incorporation of organic agriculture in the conventional food supply system, diluting the original character of organics as a form of resistance, thus resulting in the 'conventionalization of organic farming'. Conventionalizing or mainstreaming organics means that such food is produced on (large) farms that follow industrial and commercial conventions rooted in efficiency, standardization and price competitiveness and is sold via regular distribution channels, including multinational traders and processing companies, as well as supermarkets. This trend would create pressure towards reductionism and the levelling down of standards when applied by less committed actors. Marketplace dynamics 'tend to steer organic agriculture in the direction of farming systems that are technologically expedient and economically efficient in order to maximize productivity and profit, a path that is anathema to the original organic ideal' (Allen and Kovach, 2000, p225). Goodman (2000) and Guthman (2004) witness a trend (particularly in California) where large enterprises employing non-unionized, casual, migrant wage labour follow a minimalist approach to organic agriculture, adhering only to the material list of allowed inputs. At the same time, the formalization of organics in nationwide standards and the high costs involved in certification put additional pressure on small farmers (see also Klintman and Boström, 2004). This is considered antithetical to socially transformative politics and alternative visions of the world expressed by the sustainable agriculture movement. Nevertheless, all farmers, including organic food producers, operate within the larger economy and must deal with the many consequences of globalization, including the pressure to produce more per hectare. On the basis of an empirical study, however, Howard (2009) concludes that the involvement of agribusiness takes place in the domain of organic-food processing rather than in its primary production. Conventionalization in organic farming is complicated because of the capital needed and the risks involved, but also because of organic certification requirements.

When trying to maintain the social-movement character of organics, some suggest distinguishing conventional organic from alternative organic. They deem that deeper changes in political, social and economic structures and relationships are necessary, or else the conventional market system will consume its real improvements. Many consider the 'real' organic farmers to be engaged not only in producing organic 'products', but in a wider organic 'process' approach, constructing 'rural–urban bridges', or alternative producer–consumer networks.

In the US, therefore, a bifurcation occurs between orthodox organic agriculture insisting on the fundamentals of the original movement and 'organic lite' (Guthman, 2004), where large-scale production of organic products for the emerging mass market is considered a promising way forward.

## Organic farming in Europe

In Europe, although organic farming takes a different position, it has also expanded considerably since the early 1980s. This fast growth inevitably led to greater involvement of the EU, and despite hesitations within the organic agriculture movement, an official regulation was introduced in 1993 (Mikkelsen and Schlüter, 2009). This Regulation (EEC 2092/91) harmonized organic food production within the EU but requires conformation from imported products as well. The initial regulation was replaced in 2009 by EC 834/2007, which entailed a more comprehensive strategy to develop the organic sector, including the introduction of a European Community-wide logo for organic products. The growth of the organic food sector has also changed the position of organic farmers in local communities (Banks and Marsden, 2001), where they are now perceived as progressive and innovative, rising to the challenges of emerging markets and taking positive steps to safeguard their futures in the face of increasingly volatile commodity markets.

## Organic farming in Africa

It is estimated that around 3 per cent of the world's certified organic land can be found in Africa, but African farmers comprise almost 20 per cent of all certified organic farmers (Lyons and Burch, 2008). Most of this certified organic land is found in East Africa, particularly Uganda, which accounts for about half of all such land in Africa (see Box 9.5). Many more farms in Africa are operated organically in practice. Only those that are export-oriented have their produce certified, however.

Agricultural policies in most African countries have not supported organic farming and sometimes even actively hinder it. There are hardly any subsidies for organic production, and very few countries have fully implemented regulations on organic agriculture. Therefore, NGOs and private companies have been the main drivers of the growth of organic food production in Africa. They have been able to bridge the gaps between international organic standards and localized knowledge and practices, and to decrease the certification costs that smallholder farmers faced through direct support or facilitating access to other financial resources.

Export opportunities make organic certification attractive to smallholder farmers in other poor countries, particularly when this is combined with fair trade labelling.[6] In Africa, smallholder farmers may even increase their yields because of their traditionally low use of chemical inputs and the immediate impacts from investments in knowledge and capacity building (Bolwig et al, 2009; Vaarst et al, 2009). Organic farms involved in production for export prove to be significantly more profitable than those involved in conventional production. Their success in export, however, is bound with producing certified organic food in contract farming systems.[7] Organics are also promoted by concerns among African farmers and communities about the recent increase in the use of agrochemicals, which has detrimental health impacts, compounded by lax legal frameworks and their failing implementation (Oosterveer et al, 2011).[8]

Moreover, modern agricultural methods based on high external inputs have not been able to realize the increase in food production that is needed to

### Box 9.5 Organic farming in Uganda

Uganda has a large potential for profitable organic farming. Although other countries may need three to seven years for the conversion from conventional to organic farming, Uganda's northern regions may take less than a year, as agriculture there is largely organic by default. Certified organic land in Uganda has expanded from 122,000 hectares in 2003 to over 200,000 in 2008, making it the largest organic sector in Africa. The main driver for this expansion is export, which has grown at an average rate of 65 per cent per annum over these years. The National Organic Agricultural Movement of Uganda (NOGAMU) has promoted organic farming, set up local 'Shop Organic' shops, facilitated the creation of a local certification company (UgoCert) and established Uganda's organic standards. NOGAMU aims at increasing incomes and improving livelihoods in Uganda through the adoption of organic agriculture.

*Source: Daily Monitor* (2008)

feed Africa's growing population (UNCTAD and UNEP, 2008). Most of the chronically hungry are smallholder farmers who produce much of what they eat themselves, are too poor to purchase inputs and are marginalized from producer markets. It would therefore be wise to opt for a strategy that starts from the concrete living conditions of these farmers. When smallholders can be sure of selling their produce against a price premium, the dissemination of low-cost farming techniques is facilitated and their uncertainty reduced. Research has shown that organic agriculture may be attractive because yields remain at least stable when converting from traditional African farming systems. Over time, when yields increase, organics may even outperform these traditional systems, while matching the more conventional, input-intensive systems. Organic farming may increase access to food for poor African rural households by producing more food per farm and by selling food surpluses at local markets.

On a more critical note, Bakewell-Stone et al (2008) conclude that under the right circumstances, certified organic agriculture could indeed meet local food requirements and provide protection and sustainable use of natural resources, but 'the organic sector is currently being led by export-oriented organic agriculture, and does not always promote agro-ecological approaches that enhance diversity of production of food crops and lead to greater livelihood benefits' (p35).

## Conclusion on organic farming

Organic farming has become an attractive option for food producers globally. It claims to provide an answer to multiple challenges and allows an integration of wider agendas. Its methods reduce environmental pressures, while its higher consumer price and potentially increased margins for producers make it attractive from an economic point of view as well. Regarding consumers, organics use both healthy lifestyles and environmental concerns as marketing levers (see Chapter 11). Organic farming is flexible enough to collaborate with commercial partners, social movements and the state simultaneously, while at the same time retaining a recognizable and differential identity across time and space. Organics can be formalized through certification, but still operate at a local level without this formality.

The organic movement has always faced the challenge of expanding the world of organic food while maintaining its fundamental principles. The most obvious way to increase the organic sector has been to engage with conventional supply chains. This success risks side-lining the more radical agenda of transforming relations between producers and consumers, while expanding opportunities for those less committed to its core principles. This dilemma will probably become only more pressing in the future, as the growth in organics continues.

## Small Farmers and International Food Trade: Blessing or Curse?

Developing countries often view exporting food as an attractive means to stimulate their economic growth and to increase their national income. Some therefore argue for liberalizing agricultural trade through reduced domestic support, less protectionism in OECD countries and greater openness on agricultural markets in developing economies. Others have seriously questioned this strategy because markets often do not perform according to economic theory; therefore, they doubt whether farmers in developing countries will actually profit from increased exports. Many developing countries are highly dependent on exporting one or just a few (mostly unprocessed) agricultural commodities, which makes them vulnerable to price slumps and crop failures. Besides, the power in these international supply chains is often on the side of the trader or retailer rather than on the side of the producers. Overall, these agricultural commodities have been victim to declining world market prices, at least until 2008 (see Chapter 2 for a more elaborate discussion on world commodity prices). This trend was the consequence of stagnating demand, combined with rising global production. In combination with persistent poverty and hunger in many developing countries, the lack of benefits from food exports gave rise to a call for food sovereignty. In the next section, we discuss perspectives on exporting food from developing countries and whether this could contribute to increasing sustainability. We then review the campaign for food sovereignty.

### Exporting food

Overall, the global production and exports of food from developing countries have been increasing over the last 20 years. Fresh fruits and vegetables are in high demand because they contribute to better nutrition and human health and are appealing to producers because they show relatively high margins. Supermarkets aim for a year-round supply of nearly all fresh food products, including those that were seasonal in the past. Producing some of these products is attractive for smallholder farmers in developing countries because there are few economies of scale in their production and these farmers are often more flexible in seasonal and labour-intensive sectors (Gioè, 2006). Producers have profited from these increased export opportunities, which have also contributed to improving quality and productivity.

The strict food quality and safety requirements in many import markets have had a positive impact on producer practices, as Okello and Okello (2010) show for the case of green-bean growers in Kenya. Kenyan farmers had already exported green beans to Europe for several decades, but when concerns grew about the indiscriminate use of pesticides during the 1990s, the EU developed precise pesticide standards, putting Kenyan exporters under pressure. In Kenya, this resulted in a reduced use of dangerous pesticides and an increased application of alternative pest-management practices.

### Box 9.6 An example of contract details in international food trade

The Dutch fresh-fruit importer Total Fruit imports white seedless grapes from India, and its contract lists numerous details regarding product requirements, shipment arrangements, packaging instructions, product prices and the documentation necessary for shipment.

The product requirements are very detailed. They specify general characteristics and organoleptic and physical qualities, and are supplemented with precise indications to categorize the grapes as to be rejected, needing improvement, meeting the attributes and exceeding the attributes. Packaging instructions indicate specific dimensions for the carton boxes, the labels on the boxes and the manner in which the boxes should be palleted.

*Source:* Total fruit (2010)

When producing for export, these smallholders still may be confronted with different problems, including absent or deteriorating infrastructures, which complicates the transport of (particularly fresh) food products and increases the costs of logistics. Furthermore, small producers risk becoming squeezed out of the market because they cannot guarantee the products' reliability or fulfil the conditions of official and, particularly, private standards, which are required for export. These conditions relate to tracking and tracing procedures and food quality and safety standards (see Box 9.6 for an example of such requirements).

Exporting fresh fruit and vegetables, or food in general, under present conditions may perpetuate the privileged interests of the global North and marginalize those of producers in the South. In particular, importing retailers seem to have acquired more power and control over this trade than producers. The downward trend in the price of agricultural commodities on the global market exerts a permanent pressure on producers to increase their productivity and accept lower margins, or otherwise they must look for other export crops. In addition, import tariffs in the more interesting markets often range from low for unprocessed products to high for the more processed ones, thereby effectively blocking developing countries' exports of more profitable food products. This may mean that producers in developing countries remain locked in as suppliers of raw agricultural commodities, which are of low value and offer low economic returns. Profiting from growing export opportunities requires producers to repeatedly upgrade their capacity to produce higher-quality items and apply improved packaging and processing technologies. Most farmers in developing countries have few opportunities to make use of this potential because appropriate institutional support for training, input provision and financing is lacking. Many countries lack adequate institutions, and, in particular, agricultural training has received little attention from governmental authorities.

Sometimes the promotion of alternative global agri-food networks is considered a promising way forward. These networks are driven by NGOs and private bodies to create a global agri-food system 'where ecological sustainability and social justice are the normal conditions of business' (ISEAL, quoted in Hatanaka, 2010, p706). Examples of such alternative global agri-food networks are fair trade (see Chapter 7) and global organics (see above). Strengthening the position of smallholder producers in such alternative global food networks demands their participation in strong farmer movements and co-operatives that can protect their political and economic interests, provide assistance for training and capacity building and assist in developing logistics and infrastructure.

## Food sovereignty

Although food security remains an important principle in international food politics, in public debates the focus recently seems to have shifted to food sovereignty (see Box 9.7). Food sovereignty was introduced into a global political debate in 1996 by the international coalition of peasant movements La Via Campesina and refers to a nation's right to define its own food, agriculture, livestock and fisheries systems, independent of the international market forces that dominate such activities in both developed and developing societies.

Food security exists, according to the FAO, 'when all people, at all times, have physical and economic access to sufficient, safe and nutritious food to meet their dietary needs and food preferences for an active and healthy life' (FAO, 2010, p8). This definition, however, does not tell where food has to come from nor how it is produced or who produces it. Hence to secure access to food for everyone, the goal of food sovereignty is introduced, and proponents argue that 'the only lasting way to eliminate hunger and reduce poverty is through local economic development' (Rosset, 2003, p1). Food sovereignty is understood as a social right of all nations to define their own food and agriculture systems

### Box 9.7 Food sovereignty

La Via Campesina's seven principles of food sovereignty are:

- Food is a basic right for everyone, and each nation should declare access to safe, nutritious and culturally appropriate food in sufficient quantity and quality, a constitutional right. The development of the primary sector should be guaranteed to ensure the concrete realization of this fundamental right.
- Agrarian reform is necessary to give landless and farming people ownership and control over the land they work.
- Protecting natural resources, particularly land, water and seeds and livestock breeds, is necessary and should be practised by the people who work the land.
- Food trade should be reorganized because food is first a source of nutrition and only secondarily an item of trade. National agricultural policies must prioritize production for domestic consumption and food self-sufficiency.
- The globalization of hunger that is the result of the growing influence of multilateral institutions (WTO, World Bank and IMF) and multinational corporations over agricultural policies should end, and food sovereignty should no longer be undermined by them.
- Social peace; everyone has the right to be free from violence, and food must not be used as a weapon.
- Democratic control should allow smallholder farmers to have direct input into formulating agricultural policies at all levels.

*Source:* La Via Campesina (2008)

in order to produce food that fits their particular, social, cultural and ecological conditions. Food sovereignty focuses on generating a political programme aimed at strengthening the position of producers, with lesser attention to providing adequate nutrition for each individual (or household).

The strategy of food sovereignty is based on the model of peasant or family farming to achieve sustainable production with locally available resources and in harmony with local culture and traditions. Peasants and family farmers are considered to rely on a long historical experience of working with local resources, and therefore are regarded as being capable of producing the optimal quantity and quality of food in a region with only few external inputs. Agricultural production should be primarily intended for family consumption and domestic markets, while export should merely serve as an option for selling the surplus. To achieve this aim, activists from La Via Campesina demand that food and agriculture be excluded from international trade agreements, such as WTO and NAFTA. They argue that these agreements only promote international food trade and restrict the autonomy of national governments to protect their domestic food sovereignty from food being dumped on their market by exporters that are heavily subsidized by their governments, such as the US and the EU.

A fundamental question is whether food sovereignty is really feasible beyond certain specific regions. The 'call for a multi-layered, individual-to-collective sovereignty introduces confusion' (Boyer, 2010, p331), because neither the basic principles nor the institutional arrangements required for realizing this aim are clear. Nevertheless, food sovereignty serves as a concrete alternative framework for agricultural policy and intensifies the conflict between globalizing and localizing food provision.

## Conclusions

In this chapter, we reviewed various strategies that all aimed at strengthening the position of (particularly smallholder) producers in food-supply chains, while addressing some urgent sustainability challenges. Despite the limitations of such a selective overview, we have nevertheless shown that multiple strategies are present, and in different configurations they combine market opportunities, technological innovations and consumer and sustainability concerns. The presence of these various options indicates that it is unhelpful to frame the debate as only a strict opposition between alternative and mainstream farming. Both types of farming evolve, and framing conventional farming as only being static might prevent alternatives to upscale and hybrids/mergers from being invented. There is no silver bullet for creating an agriculture that reduces hunger, increases sustainability and strengthens the position of its producers under conditions of globalization because multiple fits between strategies and producers' specific conditions are possible. Hence a more integrated approach is necessary.

There is also no reason to expect technological changes in agriculture to have come to a close, neither in mainstream agriculture nor in its various alternatives, and thus innovations will repeatedly become available to producers in the future. In furthering technological development, sustainability concerns should be incorporated more explicitly than has been done in the past. Side effects should be taken into account because technological or economic optimization alone is no longer a sufficient standard for judging the prospects of technological innovations. In this process, more attention should be paid to farmers' experiences and traditional practices because these can serve as useful sources for information about the possible negative consequences of particular innovations.

---

### *Take-home lessons*

- Food producers have developed multiple strategies to ensure their future as farmers, ranging from small-scale and low-technology-intensive to high-technology-intensive options, all incorporating social and ecological impacts in some manner.
- Organic agriculture is developing fast and is becoming the icon of 'sustainable agriculture'.
- Alternative producer strategies have difficulty explaining how they can solve the problems of conventional farming without becoming mainstream themselves.

---

## Notes

1 Patenting of traditional crops is not further elaborated here, but is also considered a critical issue. Patenting for use in GM technology exposes native biodiversity and traditional knowledge related to bioprospecting. The requirements of prior informed consent from indigenous and local communities and disclosure of the origin and source contained in the Convention on Biodiversity (CBD) and the WTO Trade Related Aspects of Intellectual Property Rights (TRIPS) agreement do not seem to be robust enough to prevent this (Kamau and Winter, 2009).

2 See www.ifoam.org/growing_organic/definitions/pioneers/rudolf_steiner.php (accessed 15 March 2011).

3 See www.ifoam.org (accessed 3 January, 2011).

4 Despite its efforts to formalize certification, IFOAM is aware that non-certified organics exist as well. Such informal organic agriculture exists among farmers who deliberately refuse to be certified for financial or principle reasons, or who practice the organic principles by default.

5 'Brazil is a major exporter of organic products to Europe. Around 90% of its organic foodstuffs are produced for export markets' (Albersmeier et al, 2009, p313).

6 See, for example, the guidelines established by the organic labelling organization Naturland that make fair trade obligatory (Naturland, 2009) and those by the Organic & Fairtrade Competence Centre that support organic and fair trade supply chains (see www.organicandfair.org, accessed 17 January 2011).

7 Indeed, export of organic crops from Africa to Europe is often criticized because of the energy necessary for transport which contributes to global warming (see Chapter 5).
8 Also for Viet Nam, as Hoi et al (2009) report, vegetable producers may be concerned about the toxicity of the pesticides they use once they are aware of their potential effects on human health. These farmers did not refrain from using pesticides altogether, but they were much more selective.

## Further Reading

Boomsma, M. (2008) *Sustainable Procurement from Developing Countries: Practices and Challenges for Business Support Agencies*, KIT Publishers, Amsterdam: presents concrete examples of efforts to increase sustainability by bringing producers, traders and retailers together.

McIntyre, B., Herren, H., Wakhungu, J. and Watson, R. (2009) *International Assessment of Agricultural Knowledge, Science and Technology for Development (IAASTD): Global Report*, Island Press, Washington, DC: an overview of strategies to improve the sustainability of farming.

Rosset, P. (2006) *Food is Different: Why We Must Get the WTO Out of Agriculture*, Zed Books, London and New York: addresses the role of food producers in the world of international food politics.

## References

Albersmeier, F., Schulze, H. and Spiller, A. (2009) 'Evaluation and reliability of the organic certification system: Perceptions by farmers in Latin America', *Sustainable Development*, vol 17, pp311–324

Allen, P. and Kovach, M. (2000) 'The capitalist composition of organic: The potential of markets in fulfilling the promise of organic agriculture', *Agriculture and Human Values*, vol 17, no 3, pp221–232

Badgley, C., Moghtader, J., Quintero, E., Zakem, E., Chappell, M., Avilés-Vásquez, K., Samulon, A. and Perfecto, I. (2007) 'Organic agriculture and the global food supply', *Renewable Agriculture and Food Systems*, vol 22, no 2, pp86–108

Bakewell-Stone, P., Lieblein, G. and Francis, C. (2008) 'Potentials for organic agriculture to sustain livelihoods in Tanzania', *International Journal of Agricultural Sustainability*, vol 6, no 1, pp22–36

Banks, J. and Marsden, T. (2001) 'The nature of rural development: The organic potential', *Journal of Environmental Policy & Planning*, vol 3, pp103–121

Bolwig, S., Gibbon, P. and Jones, S. (2009) 'The economics of smallholder organic contract farming in tropical Africa', *World Development*, vol 37, no 6, pp1094–1104

Borlaug, N. (2000) 'Ending world hunger: The promise of biotechnology and the threat of antiscience zealotry', *Plant Physiology*, vol 124, no 2, pp487–490

Boyer, J. (2010) 'Food security, food sovereignty, and local challenges for transnational agrarian movements: The Honduras case', *Journal of Peasant Studies*, vol 37, no 2, pp319–351

Bray, D., Sánchez, J. and Murphy, E. (2002) 'Social dimensions of organic coffee production in Mexico: Lessons for eco-labeling initiatives', *Society and Natural Resources*, vol 15, pp429–446

*Daily Monitor* (2008) 'National Organic Agricultural Movement of Uganda (NOGAMU). Celebrating the Organic Day in Uganda', *Daily Monitor*, 14 June

Dawe, D. (2002) 'The changing structure of the world rice market, 1950–2000', *Food Policy*, vol 27, no 4, pp355–370

DeSoucey, M. and Téchoueyres, I. (2009) 'Virtue and valorization: "Local food" in the United States and France', in D. Inglis and D. Gimlin (eds) *The Globalization of Food*, Berg, Oxford and New York, pp81–95

Eicher, C., Maredia, K. and Sithole-Niang, I. (2006) 'Crop biotechnology and the African farmer', *Food Policy*, vol 31, pp504–527

ETC (2008) *Who Owns Nature? Corporate Power and the Final Frontier in the Commodification of Life*, ETC Group, Ottawa

FAO (Food and Agriculture Organization of the United Nations) (2010) *The State of Food Insecurity in the World: Addressing Food Insecurity in Protracted Crises*, FAO, Rome

FoEI (Friends of the Earth International) (2010) *Who Benefits from GM Crops?*, FOEi, Amsterdam

Friedmann, H. and McNair, A. (2008) 'Whose rules rule? Contested projects to certify "local" production for distant consumers', *Journal of Agrarian Change*, vol 8, no 2/3, pp408–434

Gioè, M. (2006) 'Can horticultural production help African smallholders to escape dependence on export of tropical agricultural commodities?', *Crossroads*, vol 6, no 2, pp16–65

GMO Compass (2008) 'Field area for Bt maize decreases', www.gmo-compass.org/eng/agri_biotechnology/gmo_planting/392.gm_maize_cultivation_europe_2009.html accessed 4 March

Goodman, D. (2000) 'Organic and conventional agriculture: Materializing discourse and agro-ecological managerialism', *Agriculture and Human Values*, vol 17, pp215–219

Green, K., Harvey, M. and Mcmeekin, A. (2003) 'Transformations in food consumption and production systems', *Journal of Environmental Policy & Planning*, vol 5, no 2, pp145–163

Guthman, J. (2004) 'The trouble with "organic lite" in California: A rejoinder to the "conventionalisation" debate', *Sociologia Ruralis*, vol 44, no 3, pp301–316

Hatanaka, M. (2010) 'Governing sustainability: Examining audits and compliance in a third-party-certified organic shrimp farming project in rural Indonesia', *Local Environment*, vol 15, no 3, pp233–244

Hoi, P., Mol, A. and Oosterveer, P. (2009) 'Market governance for safe food in developing countries: The case of low-pesticide vegetables in Vietnam', *Journal of Environmental Management*, vol 91, pp380–388

Howard, P. (2009) 'Consolidation in the North American organic food processing sector, 1997 to 2007', *International Journal of Sociology of Agriculture and Food*, vol 16, no 1, pp13–30

*Indian Express* (2010) 'System to tackle organic farming expenses on cards', *India Express*, 22 September

ISAAA (International Service for the Acquisition of Agri-Biotech Applications) (2009) *Global Status of Commercialized Biotech/GM Crops*, ISAAA, Ithaca

ISAAA (2010) *Brief 42: Global Status of Commercialized Biotech/GM Crops*, ISAAA, Ithaca

Kamau, E. and Winter, G. (eds) (2009) *Genetic Resources, Traditional Knowledge and the Law: Solutions for Access and Benefit Sharing*, Earthscan, London and Sterling

Klintman, M. and Boström, M. (2004) 'Framings of science and ideology: Organic food labelling in the US and Sweden', *Environmental Politics*, vol 13, no 3, pp612–634

La Via Campesina (2008) *Food Sovereignty for Africa: A Challenge at Fingertips*, La Via Campesina, Nyeleni

Lyons, K. and Burch, D. (2008) 'Socio-economic effects of organic agriculture in Africa', paper presented at the 16th Organic World Congress, 16–20 June, Modena

McIntyre, B., Herren, H., Wakhungu, J. and Watson, R. (2009) *International Assessment of Agricultural Knowledge, Science and Technology for Development (IAASTD): Global Report*, Island Press, Washington, DC

Mikkelsen, C. and Schlüter, M. (2009) *The New EU Regulation for Organic Food and farming: (EC) No 834/2007*, IFOAM EU Group, Brussels

Morgan, K. and Murdoch, J. (2000) 'Organic vs conventional agriculture: Knowledge, power and innovation in the food chain', *Geoforum*, vol 31, pp159–173

Naturland (2009) *Naturland Fair Richtlinien*, Naturland, Gräfelfing

Okello, J. and Okello, R. (2010) 'Do EU pesticide standards promote environmentally-friendly production of fresh export vegetables in developing countries? The evidence from Kenyan green bean industry', *Environment, Development and Sustainability*, vol 12, no 3, pp341–355

Oosterveer, P., Hoi, P. and Glin, L. (2011) 'Governance and greening agro-food chains: Cases from Vietnam, Thailand and Benin', in A. Helmsing and S. Vellema (eds) *Value Chains, Social Inclusion and Economic Development: Contrasting Theories and Realities*, Routledge, Abingdon

Raney, T. (2006) 'Economic impact of transgenic crops in developing countries', *Current Opinions in Biotechnology*, vol 17, pp1–5

Raybould, A. and Quemada, H. (2010) 'Bt crops and food security in developing countries: Realised benefits, sustainable use and lowering barriers to adoption', *Food Security*, vol 2, no 3, pp247–259

Renting, H., Marsden, T. and Banks, J. (2003) 'Understanding alternative food networks: Exploring the role of short food supply chains in rural development', *Environment and Planning A*, vol 35, no 3, pp393–411

Rosset, P. (2003) 'Food sovereignty: Global rallying cry of farmer movements', *Food First Backgrounder*, vol 9, no 4, pp1–4

Sachs, W. and Santarius, T. (2007) *Slow Trade – Sound Farming: A Multilateral Framework for Sustainable Markets in Agriculture*, Heinrich Böll Foundation and MISEREOR, Aachen and Berlin

Total fruit (2010) 'Agreement and terms of contract proposal of Indian white seedless grapes', 8 January, Barendrecht

UNCTAD and UNEP (United Nations Conference on Trade and Development and United Nations Environment Programme)(2008) *Organic Agriculture and Food Security in Africa*, UNEP-UNCTAD, New York and Geneva

USDA (United States Department of Agriculture) (2009) 'News Release No. 0036.09', USDA, 4 February

Vaarst, M., Ssekyewa, C., Halberg, N., Juma, M., Walaga, C., Muwanga, M., Andreasen, L. and Dissing, A. (2009) *Organic Agriculture for Improved Food Security in Africa*, First African Organic Conference, Kampala

Willer, H. and Kilcher, L. (eds) (2010) *The World of Organic Agriculture, Statistics and Emerging Trends, 2010*, IFOAM and FiBL, Bonn and Frick

# 10

# Restructuring Food Supply: Supermarkets and Sustainability

This chapter aims to:

- understand the key role retailers play in global food provision;
- show how retailers operate in different countries, including variations with respect to sustainability;
- discuss the changing relations between retailers and small farmers in developing countries.

## Introduction

In the course of just a few decades, supermarkets have become central locations for selling and buying food. In the US and the EU, supermarkets have dominated food retail for some time, but they are also becoming increasingly dominant in developing countries and countries in transition. The spread of supermarkets results from a combination of different societal trends, including rising incomes, urbanization, increasing female participation in the labour force and a desire to emulate Western culture. The rise in supermarkets' size and number, together with globalization in food provision, has meant that supermarkets have become obligatory passage points for most food sales.

The spread of supermarkets is fundamentally changing the way in which food is provided. Supermarkets have a profound impact on all stages, including farming, food processing, retailing and trade and consumption (Traill, 2006), and thus they have acquired a central coordinating position in food-supply chains.

To analyse these changes, we first present some background information on the growth of supermarkets and their dominant role in food provision. Building on this, we review supermarket strategies in different countries, giving specific attention to how they deal with sustainability.

**Table 10.1** *Global top ten grocery retailers*

| *Rank* | *Company* | *Food sales (2007) ($ billions)* | *Country of origin* | *Countries of operation (n)** | *Stores (n)* |
|---|---|---|---|---|---|
| 1 | Walmart | 426 | US | 15 | 8451 |
| 2 | Carrefour | 152 | France | 36 | 14,215 |
| 3 | Metro Group | 104 | Germany | 32 | 2127 |
| 4 | Tesco | 98 | UK | 13 | 4835 |
| 5 | AEON | 93 | Japan | 10 | 15,743 |
| 6 | Seven & I | 86 | Japan | 4 | 26,371 |
| 7 | Schwarz (Lidl) | 82 | Germany | 24 | 9878 |
| 8 | Kroger | 81 | US | 1 | 3619 |
| 9 | Costco | 79 | US | 8 | 560 |
| 10 | Rewe Group | 78 | Germany | 14 | 13,041 |

*Source:* Retail Planet (2010); *Deloitte (2009)

## The Growth of Supermarkets

The supermarket concept was first introduced in the US in 1916, but it was only after World War II that the model spread. Supermarket growth in Western Europe took off in the 1960s, and by the 1990s, supermarkets were responsible for the large majority of grocery sales in these countries. The expansion of grocery sales via supermarkets was accompanied by an increase in scale and a concentration into a limited number of large retailing firms. Some of these companies operate globally, but most still (primarily) serve within the boundaries of their home country. By 2009, the ten largest retailing companies accounted for 40 per cent of all groceries sold by the top 100 retailers ($5.9 trillion) (see Table 10.1).

Walmart is the world's largest grocery retailer by far, operating in 15 countries with total grocery sales (including food) of $426 billion and over 2 million employees. This retailing firm is also the largest company worldwide and dwarfs the largest food-processing company, Nestlé, which had total sales of nearly $90 billion in 2007 (see Table 10.2).

Interestingly, the world's top ten food retailers include only three US-based companies, but five Europe-based ones, while it is the reverse when considering the global food industry, which is dominated by US-based corporations. In most Western European countries, grocery sales are realized via only a few (two to five) retailing companies, and some of these companies have developed a globalization strategy, whereby Carrefour is a prime example (see Box 10.1).

Retailing companies expand from their domestic markets into other countries as the result of three interlinked processes: globalization, market liberalization and pro-corporate regulation (Lawrence and Burch, 2007). Globalization is expressed in the expanding supply of more expensive and vulnerable food products, such as fresh fruits and vegetables. Large retailers are better able than smaller ones to coordinate international supply, to profit from cost-price differences and to use their brand names to attract customers in other

**Table 10.2** *Global food industry (2007)*

| | *Company* | *Country of origin* | *Food and beverage sales 2007 ($ billions)* | *Total sales ($ billions)* |
|---|---|---|---|---|
| 1 | Nestlé | Switzerland | 84 | 90 |
| 2 | PepsiCo | US | 39 | 39 |
| 3 | Kraft Foods | US | 37 | 37 |
| 4 | The Coca-Cola Company | US | 29 | 29 |
| 5 | Unilever | The Netherlands/UK | 27 | 50 |
| 6 | Tyson Foods | US | 27 | 27 |
| 7 | Cargill | US | 27 | 88 |
| 8 | Mars | US | 25 | 25 |
| 9 | Archer Daniels Midland Company | US | 24 | 44 |
| 10 | Danone | France | 20 | 20 |

*Source:* ETC Group (2008)

than their home country. Market liberalization has enabled retailing companies to merge and to branch out to unfamiliar product lines, such as finance, insurance and medicine. Pro-corporate regulation essentially meant deregulation, away from active, interventionist policies on agriculture and food, and reregulation through voluntary, private standards (Marsden et al, 2010).

Concentration in retailing is advantageous because large retailers can make use of economies of scale when negotiating with suppliers, when investing for technological innovations in electronics and software, when applying for credits and when organizing their logistics. One indication of retailers' strong position is shown in the 'supply-chain bottleneck', where just a few retailers' buying desks control the food supply in Europe (see Table 10.3).

## Corporate Retailing Power

At one time supermarkets were simply food distributors, but now they occupy increasingly central, coordinating positions in food-supply chains, shifting the balance of power from the producer side of the chain to the retailer/consumer side of the chain. Supermarkets no longer buy just what is available on the market; rather, they find themselves in a position to determine what products suppliers must deliver under what conditions. This emerging role of

### Box 10.1 Carrefour as a global retailer

Carrefour, originally a French supermarket, has become the most globalized retailer. The company operates nearly 15,000 stores, with some 490,000 employees, in more than 36 countries in Europe, Latin America and Asia, so only about one half of the company's sales are realized in France. Carrefour operates in diverse retail formats, such as hypermarkets, supermarkets, hard discount and cash and carry.

*Source:* ETC Group (2005); Aberdeen Group (2008)

**Table 10.3** *The supply-chain bottleneck in Europe (2003)*

| | |
|---|---:|
| Consumers: | 160,000,000 |
| Customers: | 89,000,000 |
| Outlets: | 170,000 |
| Supermarket formats: | 600 |
| Buying desks: | 110 |
| Manufacturers: | 8600 |
| Semi-manufacturers: | 80,000 |
| Suppliers: | 160,000 |
| Farmers/producers: | 3,200,000 |

*Source:* Vorley (2003)

retailers as an 'institutional authority' (Dixon, 2007) is 'Janus-faced' (Harvey, 2007) because it combines the power relations it maintains with consumers *downstream*, with the different (economic) actors *upstream*, and each has its own dynamics.[1]

Upstream, this retailer influence becomes manifest in the introduction of private retail brands ('home brands') that replace manufacturer ones, as well as in the setting of specific safety, quality and environmental standards for farmers and processing firms, such as GlobalGAP (see Chapter 4). Supermarkets may also initiate 'strategic partnerships' with suppliers to generate efficiencies (Lawrence and Burch, 2007). Supermarket influence on farmers is particularly visible when they supply fresh fruits and vegetables. Much farm produce, however, such as milk, grains, sugar, etc. is processed before it reaches the retail sector, and here the impact of the retail sector is organized indirectly via the processing industry.

Downstream involvement is observable in supermarkets' decisions about shop location and format, product selection, prices and promotions, affecting the accessibility and availability of foods, as well as their prices and desirability. Adopting technological and managerial changes has enabled supermarket operators to better meet consumer preferences, but also to increase their ability to 'create consumer demand, not just meet it' (Hawkes, 2008, p682). Supermarkets are also engaged in introducing new products and services, such as convenience foods ('home-ready meals') and 'eatertainment' (offering cooking classes, producing fancy magazines free-of-charge and handing out unfamiliar food products in the shop). There is little uniformity in how these changes occur, so different configurations emerge, each with its own particular spatial presence and social dynamics (Harvey, 2007; Hawkes, 2008).

Some argue that retailers' growing power is primarily intended to increase corporations' profits to the detriment of farmers' and consumers' interests (Wailes, 2004; Young, 2004; Patel, 2007). Others, however, claim that rather than simply representing a straightforward power shift in linear supply chains, power dynamics in the retailing world emerge in 'complex and shifting *networks* of interfirm relations' (Coe and Hess, 2005, p453). Corporate retail power can be employed in various ways and has multiple (i.e. instrumental, structural and discursive) dimensions (Clapp and Fuchs, 2009). The *instrumental* dimension becomes visible in the use of resources to influence decision-making processes, and this dimension can be further specified into four levels. At the broadest level, we find the ability of retailers to organize supply (through vertical integration and standards imposed upon suppliers). The second level relates to concentration in a limited number of retail companies, while the third level concerns the operation and innovation of 'ways of doing things', or the internal operation of the shop (marketing and shop formats). The fourth and final level of retailers' instrumental power deals with their role in influencing the locations of supermarket outlets, determining their attractiveness for particular categories of consumers. The *structural* dimension of retail power entails the broader influence that corporate actors have over agenda setting and implementing. For instance, in the context of globalization, corporate retailers actively engage in developing and imposing food labelling schemes and quality standards to harmonize food products worldwide. The third dimension of retail power is *discursive* and refers to the ways in which corporate actors are able to frame an issue so as to derive political legitimacy. For instance, supermarkets have contributed considerably to the development of food safety as a dominant discourse that requires harmonized control systems, which they argue can more effectively be coordinated through private companies than governments (Marsden et al, 2010).

The multiple levels and dimensions of retailer power may lead to different configurations of suppliers and end-users. This can be illustrated with the help of two opposite models for modern retailing. One is the distanced 'distributor model' (Harvey, 2007), exemplified by Walmart and Aldi, which is organized around the conventional role of the retailer-distributor. Such retailers aim to optimize economies of scale and just-in-time distribution by revolutionizing the use of information technologies, while still trying to increase their market share through price competition. Suppliers must comply with the specific requirements of these retailers or otherwise risk losing them as outlets for their produce. Suppliers' and retailers' responsibilities are kept strictly separate. The opposite model aims at deep engagement, either upstream, downstream or both, in order to establish flexible and rapid dynamics of co-innovation to secure the availability of different own-brand food items, along a wide range of different prices and qualities. This is the 'rooted retailer model' (Harvey, 2007), of which Tesco is a prime example.

## Sustainability, food and retailers

The different models of retailing, and the various configurations of suppliers and consumers, also influence whether and how retailers engage with sustainability in food provision.

Nowadays, most supermarkets have incorporated sustainability in their corporate strategies of food provision to some extent. This is most visible in the growing sales of organic and other sustainability-labelled food products.

Conventional supermarkets became involved in sustainable food provision via the sale of organic food products in the 1980s, when UK retailer chains Tesco and Sainsbury's first stocked these products. At the start, this was done in only small quantities, but 20 years later, conventional retail outlets commanded over 80 per cent of all organic food sales in the UK (Lyons, 2007). Other areas where progress has been made are energy savings within the shop and in logistics, and the reduction of food and packaging waste.

Although many retailers nowadays subscribe to the importance of food sustainability, not all supermarkets in all countries follow similar strategies. There are considerable variations in the number of sustainable food products offered, the centrality of sustainability in sales strategies, and the commitments and qualifications of shop staff in the area of sustainable food provision (Lyons, 2007). Some retailers, such as Whole Foods Markets, actively involve consumers, as they are invited to actively take part in retailer's strategy and to join a common struggle to create a more sustainable future to: 'engage with the local community, protect the environment, distribute the food of local farmers, promote employee well-being, and service customers' desire for delicious food they can feel good about' (Johnston, 2008, p248). The company states that customers can become 'advocates for whole foods'.

Others, such as Walmart, choose to operate 'behind the consumers' back'. Walmart decided in 2005 to completely reorient its strategy on sustainability. Before, the company had been heavily criticized by environmental activists for having a very weak environmental and social performance. The company revised its policy and now claims on its website that: 'Walmart's environmental goals are simple and straightforward: to be supplied 100 per cent by renewable energy; to create zero waste; and to sell products that sustain our natural resources and the environment'.[2] The company argues that about 90 per cent of the environmental impacts occur in the supply chain. Consequently, the company radically changed its business approach in several areas, including supply-chain management, logistics and relations to rural areas. This strategy does not engage consumers in any active way, but intends to serve them by providing high-quality, low-cost products in a responsible and sustainable manner.

Again, other retailers, such as the Belgium supermarket chain Colruyt, opt for an intermediate strategy (see Box 10.2). In this case, consumers are simply serviced when shopping for sustainable food in the supermarket, but they do not necessarily have to become dedicated environmental customers.

Retailers thus may adopt different strategies to increase sustainability in food provision, but most actions are primarily seen as driven by consumer concerns, by societal demands around climate change and other environmental concerns[3] or by the opportunity to capitalize on financial savings.[4] Consumers and NGOs have contributed to promoting sustainability through various ethical consumption campaigns. Brand reputation, competitor action, the desire to pre-empt legislation and direct pressure from NGOs have all played a role

## Box 10.2 Biosupermarket Colruyt

The Belgium supermarket chain, Colruyt, is a family-owned company with a substantial share of the Belgium retail market. In its informal, down-to-earth sustainability strategy, the firm announced for its biosupermarkets (five outlets) that:

- It sells over 7000 100 per cent organic food and sustainable non-food products (e.g. made of recycled material, organic textiles, cosmetics without animal testing, and products made of natural components).
- The packaging is reusable if this option is available.
- Staff are trained to offer adequate answers to consumer questions.
- Additional information on products, organic agriculture, labels and recipes is displayed on panels and columns inside the shop.

*Source:* Colruyt, www.bioplanet.be/bioplanet/index.jsp (accessed 7 October 2009)

in encouraging supermarkets to change how they operate. Still, Fuchs et al (2009) observe that most retailers opt for the minimum strategy, which includes a rather narrow understanding of sustainability. Priorities in sustainability are based on their customers' profile and primarily address issues of nutrition and safety. If taken into account at all, environmental sustainability is understood as eco-efficiency, while the size and scale of their present business operations are hardly ever challenged as potentially fundamental sources of environmental problems. Still, many European retailers consider providing consumers with products that meet consistent food quality and safety standards that go beyond the minimum legal requirements essential to building their reputation and securing their market position. Private standards are the main instrument to achieve this because they strengthen the retailer's reputation in the eyes of consumers (Fulponi, 2006). Although food safety and quality dominate these strategies, some also take social and labour standards and animal welfare and environmental criteria into account. Thus, retailer definitions, dimensions and strategies of sustainability are not homogeneous, standard and universal, but dynamic, evolving and dependent on specific contexts. These dynamics should be taken into account when trying to understand how retailers translate sustainability concerns into forms of action. Lyons (2007) distinguishes three strategies for the inclusion of sustainability in supermarket operations:

- Maximum strategy, with a large number of sustainable food products on sale, promoted on the basis of their quality, health and environmental attributes. Sustainability is used in promotion, while educational and advocacy roles are addressed.
- Basic strategy, with an intermediate number of sustainable food products on sale based on value and price, providing little support to consumers through information, education etc.
- Minimum strategy, with a small number of sustainable food products on sale, mainly non-perishable dry goods. Staff knowledge is minimal, with little or no commitment.

## Multiple Retailer Strategies

Supermarkets may potentially contribute to greening food provision, but they can do so in multiple ways, often related to country-specific dynamics. Therefore, we will compare various European countries, the US and some African and Asian countries on their particular supermarket regimes and the presence of different strategies towards sustainability.

### European dynamics

#### *UK*

The UK is distinctive for having a very large share of the market held by the top four retailers. The country shows the highest level of own-label sales of any country in the world, the highest level of competition of chilled ready-meals, and a markedly higher occurrence of once-a-week/one-stop shopping (Harvey, 2007). The top food retailers in the UK, especially Tesco and Sainsbury's, have characteristically embraced a wide price-quality differential in their own-brand product range. Within any product line, there are a number of choices available, ranging from 'economy' at the bottom, through two or three different quality- and price-differentiated products, to top-of-the-range products whose selling point is the distinctive retailer quality brand. The strategy seems quite successful, as UK consumers are very loyal to one particular retailer for the bulk of their shopping.

Another striking feature of the British configuration of food consumption, production, exchange and distribution organized around retailing is the deep engagement of retailers, both into consumer markets downstream and in production and distribution processes upstream. Over the last two decades, intermediaries such as wholesale markets have been swept away, and new distribution systems have accelerated the route from farm to store from a matter of weeks to same-day delivery in many instances, all controlled by the front-end retailer. Long-term exclusive relationships between retailers and producers facilitate the establishment of procedures for pest control, growing regime and quality control, but also for very rapid co-innovation and experimentation. This allows supermarkets to respond instantaneously to new trends among food consumers, thus more and more replacing food-processing companies.

Compared with other countries, supermarkets in the UK have more prominently taken a leading role in changes towards greening food provision. It is retailing companies that are setting environmental standards in supply chains and may raise them in response to signals emanating from consumers. For example, retailers in the UK refused to stock food products containing GM ingredients and decided to separate non-GM ingredients throughout the supply chains for their own-brand food products (Oosterveer, 2007). Another example is Tesco's carbon footprinting scheme, which was initially developed for 20 goods (2008), including several food products, and later announced for all 70,000 products stocked on the supermarket shelves. Instead of simply calculat-

## Box 10.3 The Race to the Top

In the 'Race to the Top' project, which ran from 2000 to 2003, UK retailers were actively engaged in developing common sustainability indicators for food. This project was initiated by an alliance of farming, nature-conservation, labour, animal-welfare and sustainable development organizations to track the social, environmental and ethical performance of UK supermarkets and catalyse change within the UK agri-food sector and beyond. The project intended to identify and promote best practices of supermarkets to create benchmarks for public policy, consumers, investors, retailers and campaigners. It would also provide objective data and analysis. Seven groups of indicators were identified: environment, producers, workers, communities, nature, animals and health. An advisory group of independent experts offered advice and secured quality control. Despite initial broad interest among UK retailers, their commitment disappeared by the time the first public release of results (their individual scores on the agreed-upon indicators) was due and only Co-op, Safeway and Somerfield stayed involved in the project. In the final report, the project coordinators concluded that the supermarket sector was 'consumer-oriented in the extreme [which] has reached a point at which it is in danger of crowding out the interests of some other stakeholder groups' (Fox and Vorley, 2004, pvi). Suppliers' interests and non-consumer-related social and environmental impacts, for instance, ran the risk of disappearing from these debates and strategies.

*Source:* Fox and Peterson (2004)

ing 'food miles', this label expresses the GHGs released over the life cycle of the product in terms of grams of carbon. This method was developed in collaboration with the independent Carbon Trust Foundation. UK supermarkets have furthermore acquired a central position in the distribution of organic food (Torjusen et al, 2004). These changes signal a growing trend towards greening food provision in the UK, initiated by retailers, while government authorities play a very limited role. NGOs are involved only in monitoring the actual performance of supermarkets and seek to rank them accordingly. Supermarket companies repeatedly react to these performance lists published on NGO websites by including even stricter environmental criteria, such as reducing the number of pesticides accepted in the production process more than required in official regulations (Lang and Barling, 2007). Still, UK retailers find it challenging to be completely open about their corporate sustainability strategies, as became clear in the 'Race to the Top' project (see Box 10.3).

Generating change on issues that are not automatically in line with supermarkets' perceptions of consumer desires seems challenging and therefore probably requires more robust government interventions. The UK government, however, has shown little interest in playing an important role in pressuring supermarkets to address wider sustainability issues (Young, 2004). Except for some public campaigns on climate change, the UK government has persistently opted for a hands-off approach when it comes to increasing sustainability in food retailing.

Overall, the promotion of more sustainable food provision in the UK is primarily led by retailing corporations rather than by other social actors. This has not resulted in a radical transformation of food provision, but rather, essentially a rearrangement of socio-technical characteristics of food production, trade and retail and the images constructed at the shop-floor on food quality and sustainability. The overriding orientation of this rearrangement is towards responding to consumer concerns in terms of food safety and a reduction of environmental impacts.

### *Germany*

Germany's food retailing exhibits clear contrasts. On the one hand, German supermarkets are highly oriented towards price-discounting, relying on economies of scale, and with own-label products (where they exist) concentrating on bargains. These retailers can be further divided in three groups: centralized concerns (Metro, Tengelmann, Aldi, Lidl, etc.), which also operate on an international level; groups of co-operatives, operating internationally or nationally; and, finally, smaller nationally or regionally operating branch firms (Spiller and Gerlach, 2006). On the other hand, there are also many small retail outlets promoting traditional quality, green alternatives or local produce, sometimes with strong political support from the Green or Christian Democrat parties and backed by a powerful small farmers' organization.

In the past, the German authorities were intensively involved in securing food provisioning, but in the 1980s and 1990s this engagement lost much of its effectiveness. The BSE crisis in 2001 forced the government to reflect anew on its role in food provision (Oosterveer, 2007) because there was strong public pressure to secure food safety, protect consumers and promote sustainable development. The Ministry of Agriculture became the Ministry of Consumer Protection, Food and Agriculture and introduced an official 'Bio' label, replacing a confusing number of diverse eco-labels, and defined the policy goal to increase the market-share of organic food from 3 to 20 per cent within ten years. Retailers were offered training and extension programmes to promote organic food, while market research was made available to all interested societal actors. Furthermore, model regions were selected to create examples for regional food provision based on multi-functional agriculture, and finally a quality label for conventional agriculture was introduced (Gerlach et al, 2006). The organic food market, which remained a niche market in Germany for many years with its own (mostly small-scale) specialized retailers, changed, and since the early 2000s, large retailing companies have also been supplying organic products.

The German example clearly shows how changes in food provisioning towards more sustainability can be directly encouraged by national government interventions. These national authorities, nevertheless, were forced to operate within the limits of the CAP of the EU and to recognize the independence of supermarkets in deciding whether to accept their recommendations.

### *The Netherlands*

The Netherlands is home to one of the world's largest retailers, Ahold (operating in the Netherlands as Albert Heijn supermarkets). Together with two other retailing combinations, Albert Heijn dominates the Dutch market for fresh fruits and vegetables and has collaborated even more closely over the years. The auction

## Box 10.4 NGOs putting Dutch supermarkets under pressure

The animal rights organization Wakker Dier (Animal Alert) claimed that producing white veal forces farmers to use animal-unfriendly husbandry practices: the calves suffer from anaemia as a consequence of feeding them unhealthy food to produce this quality of meat. Between September 2008 and March 2009, Wakker Dier ran a campaign to demand supermarkets end the sale of white veal, and in the end, they received commitments from nearly all supermarkets. The action was based on publicity through radio ads, accusing specific supermarket chains of selling 'animal-unfriendly' meat. This campaign was successful also because it threatened the image of particular supermarket chains.

*Source:* www.nu.nl (accessed 15 March 2009)

system, the traditional distribution method for fresh fruits and vegetables in the Netherlands, has rapidly given way to direct sales and direct contracts between producers and retailers, organized in purchasing combinations. Currently only 5 per cent of all fresh vegetables are sold through auctioning (Bunte, 2009). One important impact of this development is that over the last ten years, the margins between the buying and selling prices have increased, generating higher profits for retailers, while price rises generally more rapidly translate in higher consumer prices than decreases result in lower consumer prices.

Generally, Dutch supermarket corporations are considered laggards in greening food compared to retailers in many other Western European countries because their primary focus is price competition. The Dutch retailer Albert Heijn claims to include sustainability throughout the corporate culture, but not to bother consumers with it. The company states that Dutch consumers simply do their shopping for food, looking for quality, safety and taste; they want to pay a low price and are not interested in reading labels about the sustainability performance of particular products. The main drivers in the Netherlands for increasing the sustainability of food provision are therefore neither supermarkets nor the government, but NGOs (see Box 10.4).

In the Netherlands, sustainability dynamics within supermarkets seem to have been driven mostly by NGOs. Environmental NGOs use public media and direct pressure from consumers as their main instruments, and most supermarkets eventually respond to these demands, but this hardly generates more active engagement from their side.

## US dynamics

Although supermarkets first emerged in the US, until the early 2000s, supermarkets played a rather minor role in the (re-)structuration of food provision. The presence of a substantial body of strict public regulation complicated concentration in the retail sector, particularly between the 1950s and the mid-1990s. Therefore, during this period, the dominance of manufacturers in the food-supply chain was maintained, and it was not until the late 1990s that power

## Box 10.5 Food deserts

The term 'food desert' refers to an area where individuals or households are unable to access a healthy and adequate diet. Some urban and rural poor, or more generally people living in rural areas, do not have access to supermarkets, grocery stores or other food retailers who offer the variety of foods needed for a healthy diet (for example, fresh fruits and vegetables, whole grains, fresh dairy and meat products). Instead, these individuals or households may rely more on food retailers or fast-food restaurants that offer limited varieties of foods. In rural areas, the issue is primarily the distance people must travel to get to a store selling healthy food, while in urban areas, this mostly relates to poverty and the cost of bus travel from poor areas to good-quality supermarkets.

It is hypothesized that because of the relative lack of access to full-service grocery stores and the easier access to fast and convenience foods, food deserts can be linked to poor diets and, ultimately, to obesity and other diet-related diseases.

*Source:* www.fooddeserts.org/ (accessed 13 January 2011); USDA (2009)

shifted towards retailers, one decade after this occurred in Western Europe (Reardon and Hopkins, 2006). Consequently, the US retail sector remained nationally fragmented and was concentrated only regionally. Even today, only Walmart and Kroger are real national chains, and most other retailers are regionally organized. This situation allowed foreign retailing corporations, such as Ahold and Delhaize, to invest in such regional retailers and acquire a strong position in the US market within a short period.

Still, in the 2000s supermarkets became rapidly concentrated in the US because legal constraints were liberalized and innovations in technology and management were introduced, particularly in logistics, which directly promoted scale enlargement (Konefal et al, 2007). Market-based forms of supply were replaced by much more integrated models of supply-chain management. Walmart, in particular, revolutionized the retail industry by developing information and management systems that allowed enormous reductions in the costs of logistics because distribution centres could become smaller and the transport fleet used much more efficiently. Supermarkets expanded in floor space and in the number and diversity of products offered. As a result of these changes, the size of the average supermarket increased and the total number of supermarkets decreased. After a high of nearly 30,000 supermarkets in the US in 1977, there were less than 25,000 left in 2000 (Kaufman, 2002). One potential consequence of this process of concentration and scale enlargement is the creation of 'food deserts', where low-income households and neighbourhoods lack access to affordable and nutritious food (see Box 10.5).

In the US, the five largest supermarket chains control 28 per cent of the retail market (with Walmart having 19 per cent in 2005). The sales by the largest 20 food retailers amounted to 64 per cent of US grocery store sales in 2009, an increase from 39 per cent in 1992.[5]

In the early days, US supermarkets did not sell foods, but in the 1960s they started selling dry foods and later fresh foods as well. As in recent years, the market for food-away-from-home has grown rapidly to nearly one half of all consumer food spending; the supermarket sections of ready-to-eat foods have also increased. For many years, competitiveness in US food retail was primarily based on price competition, and supermarkets were continuously innovating to reduce transaction costs. Since the early 2000s, comparable to the European example, the US has witnessed growing competition through quality and market differentiation. In this process, retailers also use private food quality and safety standards, but, unlike their European counterparts, they hardly rely on third-party certification and instead develop their own private schemes. US supermarkets also do not produce separate environmental reports like many European retailing corporations do, and they often do not even include a section on their environmental performance in their general annual reports (Iles, 2007). If they do publically pay attention to sustainability, US retailers mostly rely on broad general statements and concrete examples, rather than on concrete targets, statistics and trend analyses. Overall, sustainability still seems to play only a minor part in the everyday operations of most US supermarkets, except for those who have taken this up as part of their corporate identity.

## Supermarkets in the global South

Since the 1990s, many countries in the southern hemisphere have also witnessed a growing number of supermarkets in their larger cities (Boselie et al, 2003; Reardon and Hopkins, 2006; Minten, 2008). This introduction of supermarkets from domestic and foreign origins has had considerable impacts, first through the introduction of this particular retail format next to traditional food retail characterized by 'wet markets', and second because these supermarkets had to organize their sourcing of fresh fruits and vegetables, which thereby changed the lives of local farmers.

This growth of supermarkets has been rather uneven among different continents, as well as within these continents and even within countries. Still, a broad trend can be identified: in the mid-1990s, first countries in Latin America witnessed the rise of supermarkets, followed by Asia, while Africa engaged in the process only a decade later. Still today, although growing, the number of supermarkets in Africa is rather small, except for in South Africa, where the situation is comparable to that of most OECD countries (see Box 10.6 for the example of Kenya, which belongs to the more advanced countries in supermarket retailing in Africa).

The growth of supermarkets in poorer economies was fuelled by dynamics similar to those in richer countries, including urbanization, increased participation of women in the paid workforce, a growing middle class and improving infrastructures, but particularly by pronounced population growth. The rise of supermarkets in Latin America and Asia was also the result of the liberalization of direct foreign investment in many countries in the 1990s, which enabled large foreign retail corporations to invest there. The saturation and heavy competition in their domestic markets drove these companies to search for foreign markets to increase their margins. More prosperous consumers in

## Box 10.6 Kenya's retail sector

The retail food sector in Kenya has expanded significantly in recent years as a result of the country's rapid population growth and urbanization, a growing middle class and a financially attractive business environment. In 2010, the supermarket sector represented one third of the total retail space, with nearly 300 hyper- or supermarkets and many more convenience stores in both large cities and small towns. The growth in this form of retailing meant that nearly as many fresh fruits and vegetables produced in Kenya are sold through supermarkets as are exported. Not only wealthy and middle-class consumers shop at these retailers but poor consumers do as well, albeit less frequently and spending less.

Kenya's supermarket sector is the most advanced in East Africa, and some of its companies, such as Nakumatt and Uchumi, are investing in neighbouring countries: Uganda, Tanzania and Rwanda. Through such investments these countries are establishing a modern food distribution system.

*Sources:* Reardon et al (2003); USDA (2010)

non-OECD countries find supermarkets attractive because their prices are lower and convenience is higher, while they claim to sell safe food of better quality (Reardon and Hopkins, 2006). In addition, supermarkets may even be more attractive for them when they offer credit cards and other financial services.

In general, most supermarkets initially provide only processed and packaged foods, such as noodles, milk, drinks and grains. They tend to sell fresh foods later, when they are able to organize their supply chain and compete with traditional (wet) markets. Supermarket procurement has changed the roles of producers from their roles in traditional food markets. Modern retailers require higher and more consistent quality, year-round supply, larger volumes, adherence to higher safety standards, more stringent payment terms, etc. In the early phases of supermarket diffusion, two different retail segments are formed. The first is a small set of leading chains acquiring a small share of the food market and modernizing their procurement systems to clarify the specific requirements and incentives that producers were facing. Next to this group, a large set of second-tier chains and traditional retailers is formed that controls a large segment of the food market and manifests traditional food market conditions and requirements. As supermarket diffusion continues, the situation reverses and farmers face a food market dominated by the leading supermarket chains (Viteri, 2010). In nearly all countries where supermarkets have been introduced, small shops disappeared rapidly and in large numbers (Humphrey, 2007), although such shops remain relevant in the sale of fresh fruits and vegetables (see Box 10.7 for the example of Viet Nam, which deviates from this general trend).

Consumers find buying food in supermarkets attractive because as some studies (e.g. D'Haese and Van Huylenbroeck, 2005) report, food prices on average are higher in local shops than in supermarkets.[6] Other considerations are the better-quality food (image) that supermarkets offer, in combination with the possibility of one-stop shopping and the presence of a clean shopping environment.

### Box 10.7 Slow penetration of supermarkets in urban Viet Nam

Traditionally, fresh-food marketing in Viet Nam involved supply networks operating through many wholesale and retail markets scattered throughout the larger cities. In the mid-1990s, supermarkets of both domestic and foreign origin were introduced, but ten years later, they still cover only minimal percentages of fresh fruit and vegetable sales (Cadilhon et al (2006) report a market share of only 2 per cent). In 2004, in Hanoi around 29,211 traders worked in traditional wholesale and retail markets, while the modern retail sector employed 1917 people. This unexpected development can be explained by the importance that Vietnamese consumers attach to fresh-quality produce and the lack of guarantees that the more anonymous supermarkets can offer for the quality and safety of fresh fruits and vegetables.

*Source:* Cadilhon et al (2006); Moustier (2006); Hoi (2010)

## Supermarkets and Fresh Fruit Exports

Exporting fresh fruits and vegetables to supermarkets in the global North offers interesting income opportunities for farmers in developing countries. Since the 1990s, fresh fruits and vegetables have become a key area of competition among supermarkets as one of the few growth opportunities in food sales (Dolan and Humphrey, 2000). Supermarkets try to access their supply against the lowest possible price, but they also compete on the basis of variety and packaging in addition to quality and safety, which places considerable pressure on farmers to comply with all these supermarket requirements. Retailers want to assure year-round availability of all fresh fruits and vegetables, and therefore, they have to source them from different regions, while guaranteeing the same quality, variety and packaging. Securing quality and safety was initially organized through explicit coordination along the supply chain from the retailing company, but since the mid-1990s, this primarily has been achieved through independent certification. Initially, the preferred certification scheme was HACCP (see Chapter 4) for food safety, later supplemented with management standards, such as ISO 9000 for quality management, ISO 14000 for environmental protection and SA 8000 for social performance. Over time, many of these have more or less merged in GlobalGAP (see Chapter 4). When promoting such certification schemes, most supermarkets try to force farmers to cover the bulk of the additional costs involved. Supermarkets have an interest in supporting third-party certification instead of imposing their own corporation-specific standards because it enhances public recognition and allows for broader application of similar standards in the market, which contributes to a more level playing field with their competitors. Still, some retailers, particularly in the US, prefer to develop and use their own company standards and thereby strengthen their competitive power.

Over time, the following four pillars have emerged as essential for contemporary supermarket procurement (Reardon et al, 2003):

- centralization of procurement by chain;
- shifting from a reliance on spot markets (in particular traditional wholesale markets) towards growing use of specially committed wholesalers to cut coordination costs and enforce private standards and contracts on behalf of supermarkets;
- supermarkets' use through their dedicated wholesalers lists of preferred suppliers, which function as informal but effective contracts;
- use of private quality and safety standards.

Retailers increasingly structure their supply chain by selecting a group of preferred suppliers, and although these suppliers still have to carry most of the risks in production, they have a guaranteed outlet for their produce. The position of producers that are not included in this arrangement is becoming more vulnerable because such producers depend on the spot market, where prices may vary enormously. Overall, it seems, however, that small-scale farmers are not necessarily excluded from selling to supermarket chains, provided they can meet their requirements in terms of quality, safety, and production and delivery details. In the domain of fresh fruits and vegetables in particular, better-equipped small-scale producers may even have a comparative advantage over larger growers through their flexibility, specialization and quality assurance. Besides, supermarket chains may consider sourcing from small-scale producers an attractive way to display their socially responsible business practices and a welcome means to diversify their suppliers and not be completely dependent on only a few large ones. Retailers may even assist their smaller suppliers with inputs and assistance to increase their capacity and assure the quality and on-time delivery of their produce. This is particularly helpful in countries where public services are not available to support producers. Nonetheless, supermarket supply chains are much more streamlined than traditional market channels, which reduces the number of suppliers and wholesalers involved and necessarily advances the position of larger growers (Boselie et al, 2003). Larger suppliers can more easily comply with retailer requirements, but they are also better capable of advancing their interests through negotiations. These possibilities could also be available to smallholder farmers, provided they organize themselves in collectives and do not operate individually. In the context of fresh fruit and vegetable production and export, multiple strategies are available for growers (Jaffee and Masakure, 2005), as long as they remain prepared to adapt to changes on the market and to the way in which global supply chains are organized. Those that are unable to fulfil the necessary conditions are left with only one other option: to sell on the informal local market and to accept more uncertainty and often a lower price.

## Corporate Social Responsibility

In the last decade, the concept of CSR has received increasing attention, and today many view it as an important additional driver in changing supermar-

## Box 10.8 Retail environmental sustainability code

The voluntary retail environmental sustainability code, developed by the European Retail Round Table, aims to promote sustainable production and consumption. Therefore, signatories to this code (such as Tesco and Carrefour) commit to implementing measures for reducing the environmental footprint of their operations in:

- *sourcing:* by promoting more environmentally sustainable sourcing of products;
- *resource efficiency:* by improving the environmental performance of the retailers' premises;
- *transport and distribution:* by improving the environmental performance of distribution;
- *waste management:* by putting in practice measures aiming to prevent or reduce the impact of waste on the environment;
- *communication:* by improving customer-oriented communications that encourage more sustainable consumption;
- *reporting:* by reporting regularly on the commitments above.

*Source:* EuroCommerce and European Retail Round Table (2010)

ket practices. Different food retailers, as well as many other private companies, claim they want to make a positive contribution to society. Although precise definitions of CSR are lacking, retailers' ambition is based on the understanding that society and private business are not distinct entities and that responsibility for society and the environment belongs to the core of doing business (Jones et al, 2006). It entails 'a process whereby a company assumes responsibility, across its entire supply chain, for the social, ecological and economic consequences of the company's activities, reports on these consequences, and constructively engages with stakeholders'.[7] CSR is sometimes shortened as 'People, Planet and Profit', or 'the triple bottom line', and this intention is expressed through special reports from private companies on their 'CSR-performance'. In addition to food-safety requirements, CSR-involved retailers have often also imposed minimum social standards on themselves, their member companies and their suppliers. The presence of such standards and their required implementation may increase awareness about social aspects and improve performance among the different producers and companies involved throughout the supply chain (for an example of a retail sustainability code, see Box 10.8).

The consultancy firm, the Aberdeen Group (2008), suggests that companies enhancing their overall corporate responsibility reap impressive gains in competitive advantage via lower costs, as well as remarkable related advantages, such as customer loyalty, new customer acquisition and customer retention and satisfaction. Cost reductions resulting from CSR are immediately reflected in yearly financial reports, but the other impacts are less visible. Companies are trying to capitalize on such non-financial results through annual reports, websites and other forms of publicity. One often-applied tool to objectify the retailing company's environmental performance is certification, for instance, through ISO 14001. Compared to many other retail sectors, the food industry seems to be relatively well respected by consumers for its CSR performance.[8]

CSR initiatives are criticized for the absence of mechanisms that allow external partners to influence the (selection of) criteria and aims that retailers set for themselves. Because retailers set their own standards, small producers may have difficulty complying with them; in particular, they may find it difficult to meet the costs of the investments required, including adapting administrative systems. Unless they request external monitoring of their performance, there is also little opportunity to verify whether retailers have actually performed according to their set aims. Therefore, many consider the standards being set at the minimum by corporations as tending towards 'greenwash' and not towards real sustainable change. A final comment concerns CSR's intention to fill in the voids in official regulations; by developing concrete alternatives, private companies compete with official governmental interventions, which thereby risk becoming weakened.

## Conclusion

This chapter illustrated supermarkets' central position in contemporary food provision, including in international trade and with high impacts on the eventual promotion of sustainability. Different contexts and strategies were presented to show the variations and complexities involved, also with regard to the relationships between primary producers and retailers.

Globalization strengthens retailers' central position because they are directly influencing how the global food supply is organized. In the process, they may be confronted with the tension between further outsourcing tasks to reduce fixed costs and to promote their own flexibility and losing control over the food supplied. This loss of control may be problematic because consumers still hold retailers accountable for food quality and safety. Retailers must balance flexibility and control when considering their role in food-supply chains. For this reason, supermarkets are among the principal drivers of private standards, which have taken over the steering of producer practices from national laws and domestic regulations.

Supermarkets may have different reasons for increasing, or not increasing, the sustainability of their food-provision strategies. Retailing corporations oriented towards providing food as cheaply as possible may be less interested in doing more than the law requires. Increased attention to sustainability concerns may pay off financially, however, by reducing costs or by attracting customers who are willing to pay extra for food at the higher (sustainability) end of the product range. By focusing on the higher end of the consumer market, where demands for more quality foods, including demands for increased sustainability, are strong, retailers may profit from higher margins. Such economic considerations, however, cannot explain all transformations because non-economic reflections come into play as well, such as the intention to offset NGO or consumer protests and preserve the company's positive image. Other changes are made in reaction to concerns about sustainability and the future provision of food, as expressed within the company.

Involving supermarkets in sustainability transitions is essential because, for the foreseeable future, supermarkets will remain key actors in permanent transformations in the contemporary food supply through globalization, rapid urbanization, technological innovation and consumer demands for healthy and varied food products. Potential contributions from retailers to further the greening of food provision should be taken more seriously than many are at present. These contributions should not be approached in isolation, but as a key node in networks connecting primary food production with final consumption and disposal. This is where understandings and practices related to sustainable food are (re)created, because the supermarket floor constitutes the location where the system of food provision meets the consumer. As the 'Race to the Top' project in the UK showed, supermarkets may actively contribute to more sustainability, but they primarily react to consumers and their interpretation of sustainability, which may evolve over time but seems to focus more on issues such as food safety, sustainability and animal welfare than on the social consequences for producers.

Strengthening large retailers' contribution to greening food provision may occur via different trajectories: provider induced, government induced and consumer induced. Which trajectory will be the most promising depends largely on the specific national culture and political developments and on the specific local or national context of the organization of the retail food sector, as the divergences in supermarket characteristics among countries and regions remain surprisingly large.

Within this context, the shifting balance of power between supermarkets and governments must be acknowledged. Different large retailing corporations operate beyond the control of national authorities, so forcing them to implement sustainability measures for public-interest reasons cannot be done by individual governments but requires coordinated efforts from multiple societal actors.

---

## *Take-home lessons*

- Supermarkets have become the key node in global food provision.
- Many large retailers include sustainability concerns in their business practices, to a greater or lesser extent, mostly under pressure from governments and NGOs or to strengthen their marketing position.
- Promoting food sustainability by retailers may improve the performance of small producers in developing countries, but may also force them to take up much of the costs involved in adapting their practices.

---

## Notes

1. Upstream refers to the suppliers of food products and includes farmers, processors and traders, while downstream entails consumers as the end-users of the food products sold through supermarkets.
2. See www.walmartstores.com (accessed 11 February 2009).
3. The organization of present-day food provision in OECD countries is increasingly affected by 'ethical values circulating between spheres of consumption, retail and production, as well as civil society and the media' (Hughes et al, 2008, p363).
4. Retailers may adopt environmental standards to attract concerned customers and/or to reduce costs, for example, by diminishing energy use, which has a considerable economic impact (Iles, 2007).
5. See www.ers.usda.gov/Briefing/FoodMarketingSystem/foodretailing.htm (accessed 6 January 2011).
6. This observation contradicts Minten's (2008) conclusion that food prices in supermarkets are consistently higher than in the traditional retail markets.
7. See http://mvoplatform.nl/what-is-csr (accessed 6 January 2011).
8. See Food-navigator-USA (31 March 2010), www.foodnavigator-usa.com/Financial-Industry/Food-industry-well-respected-for-CSR-efforts-Survey (accessed 2 April 2010).

## Further Reading

Burch, D. and Lawrence, G. (eds) (2007) *Supermarkets and Agri-food Supply Chains: Transformations in the Production and Consumption of Foods*, Edward Elgar, Cheltenham and Northampton: discusses the power of supermarkets in different continents, including in developing countries.

McCullough, E., Pingali, P. and Stamoulis, K. (eds) (2008) *The Transformation of Agri-Food Systems: Globalization, Supply Chains and Smallholder Farmers*, Earthscan, London: presents case studies on the impacts of global supermarket supply chains on small-scale farmers in developing countries.

## References


Aberdeen Group (2008) *Sustaining the Global Food Supply Chain*, Aberdeen Group, Boston

Boselie, D., Henson, S. and Weatherspoon, D. (2003) 'Supermarket procurement practices in developing countries: Redefining the roles of the public and private sectors', *American Journal of Agricultural Economics*, vol 85, no 5, pp1155–1161

Bunte, F. (2009) *Prijsvorming glastuinbouw*, LEI, The Hague

Cadilhon, J., Moustier, P., Poole, N., Tam, P. and Fearne, A. (2006) 'Traditional vs. modern food systems? Insights from vegetable supply chains to Ho CHi Minh City (Vietnam)', *Development Policy Review*, vol 24, no 1, pp31–49

Clapp, J. and Fuchs, D. (eds) (2009) *Corporate Power in Global Agrifood Governance*, The MIT Press, Cambridge, MA

Coe, N. and Hess, M. (2005) 'The internationalization of retailing: Implications for supply network restructuring in East Asia and Eastern Europe', *Journal of Economic Geography*, vol 5, pp449–473

Deloitte (2009) *Emerging from the Downturn: Global Powers of Retailing 2010*, Deloitte Touche Tohmatsu, London

D'Haese, M. and Van Huylenbroeck, G. (2005) 'The rise of supermarkets and changing expenditure patterns of poor rural households: Case study in the Transkei area, South Africa', *Food Policy*, vol 30, pp97–113

Dixon, J. (2007) 'Supermarkets as new food authorities', in D. Burch and G. Lawrence (eds) *Supermarkets and Agri-food Supply Chains: Transformations in the Production and Consumption of Foods*, Edward Elgar, Cheltenham and Northampton, pp29–50

Dolan, C. and Humphrey, J. (2000) 'Governance and trade in fresh vegetables: The impact of UK supermarkets on the African horticulture industry', *Journal of Development Studies*, vol 37, no 2, pp147–176

ETC (2005) *Oligopoly, Inc. 2005: Concentration in Corporate Power*, ETC Group, Ottawa

ETC (2008) *Who Owns Nature? Corporate Power and the Final Frontier in the Commodification of Life*, ETC Group, Ottawa

EuroCommerce and European Retail Round Table (2010) *Retail Environmental Sustainability Code*, Eurocommerce and ERRT, Brussels

Fox, J. and Peterson, H. (2004) 'Risks and implications of bovine spongiform encephalopathy for the US: Insights from other countries', *Food Policy*, vol 29, pp45–60

Fox, T. and Vorley, B. (2004) *Stakeholder Accountability in the UK Supermarket Sector: Final Report of the 'Race to the Top' Project*, IIED, London

Fuchs, D., Kalfagianni, A. and Arentsen, M. (2009) 'Retail power, private standards, and sustainability in the global food system', in J. Clapp and D. Fuchs (eds) *Corporate Power in Global Agrifood Governance*, The MIT Press, Cambridge, MA and London, pp29–60

Fulponi, L. (2006) 'Private voluntary standards in the food system: The perspective of major food retailers in OECD countries', *Food Policy*, vol 31, pp1–13

Gerlach, S., Kropp, C., Spiller, A. and Ulmer, H. (2006) 'Die Agrarwende-Neustrukturierung eines Politikfelds', in K. Brand (ed) *Von der Agrarwende zur Konsumwende? Die Kettenperspektive; Ergebnisband 2*, Oekom Verlag, Munchen, pp37–61

Harvey, M. (2007) 'The rise of supermarkets and asymmetries of economic power', in B. David and G. Lawrence (eds) *Supermarkets and Agri-food Supply Chains; Transformations in the Production and Consumption of Foods*, Edward Elgar, Cheltenham and Northampton, pp51–73

Hawkes, C. (2008) 'Dietary implications of supermarket development: A global perspective', *Development Policy Review*, vol 26, no 6, pp657–692

Hoi, P. (2010) *Governing Pesticide Use in Vegetable Production in Vietnam*, WUR, Wageningen

Hughes, A., Wrigley, N. and Buttle, M. (2008) 'Global production networks, ethical campaigning, and the embeddedness of responsible governance', *Journal of Economic Geography*, vol 8, pp345–367

Humphrey, J. (2007) 'The supermarket revolution in developing countries: Tidal wave or tough competitive struggle?', *Journal of Economic Geography*, vol 7, pp433–450

Iles, A. (2007) 'Seeing sustainability in business operations: US and British food retailer experiments with accountability', *Business Strategy and the Environment*, vol 16, pp290–301

Jaffee, S. and Masakure, O. (2005) 'Strategic use of private standards to enhance international competitiveness: Vegetable exports from Kenya and elsewhere', *Food Policy*, vol 30, pp316–333

Johnston, J. (2008) 'The citizen-consumer hybrid: Ideological tensions and the case of Whole Foods Market', *Theory and Society*, vol 37, pp229–270

Jones, P., Comfort, D., Hillier, D. and Eastwood, I. (2006) 'Corporate social responsibility: A case study of the UK's leading food retailers', *British Food Journal*, vol 107, no 6, pp423–435

Kaufman, P. (2002) *Food Retailing*, USDA, Washington, DC

Konefal, J., Bain, C., Mascarenhas, M. and Busch, L. (2007) 'Supermarkets and supply chains in North America', in D. Burch and G. Lawrence (eds) *Supermarkets and Agri-food Supply Chains: Transformations in the Production and Consumption of Foods*, Edward Elgar, Cheltenham and Northampton, pp268–288

Lang, T. and Barling, D. (2007) 'The environmental impact of supermarkets: Mapping the terrain and the policy problems in the UK', in D. Burch and G. Lawrence (eds) *Supermarkets and Agri-food Supply Chains: Transformations in the Production and Consumption of Foods*, Edward Elgar, Cheltenham and Northampton, pp192–215

Lawrence, G. and Burch, D. (2007) 'Understanding supermarkets and agri-food supply chains', in D. Burch and G. Lawrence (eds) *Supermarkets and Agri-food Supply Chains: Transformations in the Production and Consumption of Foods*, Edward Elgar, Cheltenham and Northampton, pp1–26

Lyons, K. (2007) 'Supermarkets as organic retailers: Impacts for the Australian organic sector', in D. Burch and G. Lawrence (eds) *Supermarkets and Agrifood Supply Chains: Transformations in the Production and Consumption of Foods*, Edward Elgar, Cheltenham and Northampton, pp154–172

Marsden, T., Lee, R., Flynn, A. and Thankappan, S. (2010) *The New Regulation and Governance of Food: Beyond the Food Crisis?*, Routledge, New York and London

Minten, B. (2008) 'The food retail revolution in poor countries: Is it coming or is it over?', *Economic Development and Cultural Change*, vol 56, no 4, pp767–789

Moustier, P. (2006) 'Summary of main findings from literature and study on supermarket development', in P. Moustier, T. Phan, H. An, V. Binh, M. Figuié, N. Loc and P. Tam (eds) *Supermarkets and the Poor in Vietnam*, M4P and CIRAD, Hanoi, pp30–41

Oosterveer, P. (2007) *Global Governance of Food Production and Consumption: Issues and Challenges*, Edward Elgar, Cheltenham and Northampton

Patel, R. (2007) *Stuffed and Starved: Markets, Power and the Hidden Battle for the World Food System*, Portobello Books, London

Reardon, T. and Hopkins, R. (2006) 'The supermarket revolution in developing countries: Policies to address emerging tensions among supermarkets, suppliers and traditional retailers', *The European Journal of Development Research*, vol 18, no 4, pp522–545

Reardon, T., Timmer, P., Barrett, C. and Berdegue, J. (2003) 'The rise of supermarkets in Africa, Asia, and Latin America', *American Journal of Agricultural Economics*, vol 85, no 5, pp1140–1146

Retail Planet (2010) 'Top 20: Tesco poised for rapid growth', press release 29 June, www.planetretail.net

Spiller, A. and Gerlach, S. (2006) 'Wertschöpfungsketten für Bio-Produkte: Getrennte Welten', in K. Brand (ed) *Von der Agrarwende zur Agarwende zur Konsumwende? Die Kettenperspektive*, Oekom Verlag, München, pp83–106

Torjusen, H., Sangstad, L., Jensen, K. and Kjaerness, U. (2004) *European Consumers' Conceptions of Organic Food: A Review of Available Research*, SIFO, Oslo

Traill, W. (2006) 'The rapid rise of supermarkets?', *Development Policy Review*, vol 24, no 2, pp163–174

USDA (United States Department of Agriculture) (2009) *Access to Affordable and Nutritious Food: Measuring and Understanding Food Deserts and Their Consequences*, USDA, Washington, DC

USDA (2010) *Kenya, 2010 Retail Food Sector Report*, USDA, Washington, DC

Viteri, L. (2010) *Fresh Fruit and Vegetables: A World of Multiple Interactions*, WUR, Wageningen

Vorley, B. (2003) *Food Inc.: Corporate Concentration from Farm to Consumer*, UK Food Group, London

Wailes, E. (2004) 'Rice: Global trade, protectionist policies, and the impact of trade liberalization', in M. Aksoy and J. Beghin (eds) *Global Agricultural Trade and Developing Countries*, World Bank, Washington, DC, pp177–193

Young, W. (2004) *Sold Out: The True Cost of Supermarket Shopping*, Vision, London

# 11

# Consumer Involvement in Sustainable Food Provision

The chapter aims to:

- discuss the changing roles of consumers under global modernity;
- present a social-practices approach to analysing consumer behaviour;
- introduce different consumer strategies towards sustainable food provision.

## Introduction

Consuming food involves daily routines, cultural practices and economic calculations, and thus consumer involvement in sustainable food provision is filled with complications. For instance, research consistently shows that consumers in developed economies express positive attitudes towards buying sustainable food, but only a minority actually buys it. So, why do consumers not behave according to their expressed values? Some suggestions are that consumers are unwilling to change their daily routines, are manipulated through marketing and by large food retailers, do not trust the 'alternatives' or are unwilling to pay higher prices for more sustainable food products. It is difficult to determine which of these explanations is correct. This is just one of the issues related to the links between consumers and sustainable food. Still, analysing food consumption from a sociological perspective is relatively new.

Until the 1990s, sociologists largely ignored (food) consumption as a relevant field of study, and mostly left it to nutritionists, economists and psychologists. Nowadays, however, many social scientists analyse (food) consumption and consumer practices, conceptualizing consumers as social agents who are actively engaged in shaping essential social practices. Consuming food entails a broad array of activities, such as buying products and services; transporting, preparing and eating them; and finally disposing of or recycling the remaining waste. These different social practices display large variations and have changed considerably over time through rising incomes, cultural developments, increased communication and international travel, changing household composition, division of labour, etc.

As a subject for environmental policy, consumption emerged broadly only in the 1990s, when it was included in Agenda 21 during UNCED (in Rio de Janeiro in 1992).[1] Since the start of the second millennium, environmental social scientists have also examined consumer practices and helped develop strategies for more sustainable consumption (see for instance Shove, 2003; Spaargaren, 2003).

In this chapter, we start by sketching the place food consumption occupies in broader social theory and some of the key analytical frameworks that are used to study it. In the next section, a conceptual tool based on a structuration perspective is presented, which we apply to the case of organic food consumption and dietary change. In the concluding section, we reflect on the possible roles consumers can play in promoting more sustainable food provision.

## Food Consumption Under Global Modernity

Food consumption has drawn social scientists' attention and over the years, topics such as food cultures, food security and human health have been studied extensively. Recent studies have shown that conventional eating habits in richer economies have radically changed since the 1970s (Counihan and Esterik, 1997). Unfamiliar fruits and vegetables have become everyday goods and global sourcing a common practice (Millstone and Lang, 2003). Exotic recipes became ordinary dishes (James, 1996), and eating out in restaurants and canteens regular events (Warde and Martens, 2000), though buying food for home consumption remains important (see Box 11.1 for an example of present-day US food consumption).

Over the past few decades, especially in OECD and transition countries, two trends in food consumption are visible: with rising incomes, consumers spend a lower percentage of their earnings on food and they consume more meat. As most consumers can satisfy their basic nutritional needs, the demand

### Box 11.1 US food consumption statistics

Americans spend about one tenth of their income on food, with consumption patterns indicating unsustainability, in several respects:

- The average American consumes 2775 calories per day (2007), up 28 per cent since 1970.
- The average American consumes nearly 91kg of meat per year (2007), an increase of more than 10 per cent from 1970.
- Of all US adults, 67 per cent are considered overweight or obese (2006).
- More than one quarter of all edible food is wasted at the consumer level.

*Source:* Center for Sustainable Systems (2009)

for new and more diverse products is growing. For instance, as Van Otterloo (2000) explains in the case of the Netherlands, until the early 20th century, accessing sufficient food dominated most individuals' daily activities. Within a few decades, the situation changed into one of abundance.[2] Rising incomes in the Netherlands in the second half of the 20th century favoured foods that were more expensive, more processed and contained different basic ingredients. Supply chains had become longer, and large-scale industries further differentiated and produced in increasing quantities. By the end of the second millennium, nearly all Dutch citizens had access to more than 15,000 different food products and consumed more meat, cheese and fresh fruits, and less milk and potatoes.

Comparable changes have occurred in the other European countries, including the UK (Cheng et al, 2007) and Belgium (Mestdag, 2005). In the mid-20th century, food seemed to become increasingly standardized through mass production and consumption of uniform food products manufactured by a well-developed food industry. This phase of simple modernization was based on the production of homogeneous and durable food products that were open to trade and canned, dried or frozen for long shelf life. Since the 1990s, this trend has become less prominent as, although food provision has remained large scale and partly industrialized, it also has become geared to growing diversity in consumer demand and retailer strategies in contemporary societies. Patterns of food production and consumption have had to accommodate flexibility in production methods and distinctions in individual tastes (Fine et al, 1996). These trends may show similarities among different countries as they are related to some broad trends, such as globalization, urbanization, rising incomes and changing gender relations. The impacts of such general trends differ considerably, however, among the different locations and even among the different households within one region. Considerable variation also exists in knowledge about food production, processing and trade, and about cooking[3] and recycling waste among and within the households, which is partly related to gender, age, income and social class.

Stretching the distances in time and space between food production and consumption and intensifying the application of science and technology, which is characteristic of globalizing food provision, gave rise to new consumer concerns. These changes call for a better understanding of food consumption.

## Food Consumption in Social Theory

Social scientists have developed several different arguments related to how food consumption practices may evolve towards greater sustainability (Jackson, 2005; see also Miller, 1995; Tansey and Worsley, 1995; Howes, 1996; Atkins and Bowler, 2001; Sassatelli, 2007; Shove et al, 2009). Three lines of argument can be distinguished in this literature: those focusing on (1) individual consumers, (2) the supply system and (3) social consumption practices.

## Individual food consumers as rational actors

Some economists and psychologists consider consumers as rational actors and try to establish the relationship between individual attitudes and purchasing behaviours. In the case of food consumption, such social scientists approach the individual as the main intervention level because consumers determine the fate of green products and practices. Focusing on the discrepancy between attitudes and behaviours (Padel and Foster, 2005), they attempt to explain the limited consumer responses to the sustainable food products that are offered.[4] According to this perspective, individuals make their (buying) decisions as rational actors. Their purchasing decisions are influenced by normative beliefs, habits and the perceived costs and benefits of a particular choice (Lunt, 1995). When analysing the slow progress in sustainable food consumption from this perspective, scholars point to the lack of concrete information available to consumers about the benefits of more sustainable food products (Vermeir and Verbeke, 2006), to the complications in buying such products in conventional retailing channels, and to the price differential between conventional and sustainable foods (Padel and Foster, 2005) (see Box 11.2 for a case that questions the singular importance of price in purchasing decisions related to sustainable food).

Consumers do not consider 'sustainable food' to be an important purchase criterion because such foods are not perceived to surpass conventional foods in taste and shelf life (two important factors in consumer choice). Furthermore, attitudes and behaviours towards organic foods seem to be more strongly related to health benefits than to perceived environmental benefits (Shepherd et al, 2005). Interestingly, within the same group of consumers, the relative importance of motives and barriers appears to vary among different product categories, whereby sustainability seems to be a more important criterion for fresh food than for processed products (Padel and Foster, 2005).

Despite the popularity of this individualist perspective on consumer practices, it has serious limitations. These limitations are visible, for instance, in the limited success of public campaigns to change consumer behaviours towards more sustainability, showing that information alone is hardly capable of changing actual behaviours. Consumer decision-making, particularly in such mundane domains as food, often occurs under time constraints and cognitive

### Box 11.2 Does price determine sales of sustainable food?

The Dutch Agricultural Economics Institute (LEI) experimented with reducing prices to see how sales of organic foods would be affected. They reduced the price of organic products in grocery stores by 40 per cent. For organic eggs, beef and cereals, lowered prices resulted in small but positive increases in food sales, but for organic potatoes, mushrooms, milk, rice and pork, lower prices did not result in significantly increased food sales. The researchers concluded that pricing is important, but not necessarily determinative, in the sales of organic food. Other considerations, such as image, also are important.

*Source:* Baltussen et al (2006)

limits, so rationality is always bounded to a rapid comparison of only a few options. Furthermore, a focus on individual behaviour obscures the relevance of social dynamics in understanding consumption behaviour. Decision-making is often done collectively and consuming does not occur in a vacuum but under conditions influenced by social structures.

## Systems of food provision

Other social scientists try to explain food consumption by focusing on structural characteristics of the current, market-based system of food provision (Seyfang, 2006). Consumers are conceptualized as passive or 'captive', and are considered to be largely dependent on the dominant food system, which is controlled by multinational food processors, big retailing companies and other large firms (Pollan, 2008; Roberts, 2009). Much actual consumption is embedded in social, cultural, economic and institutional infrastructures over which people have little influence (Barnett et al, 2011). Or, as Hobson (2002, p103) states, 'individual consumption practices (are complex and entrenched), positioned within contexts and infrastructures not conducive to living sustainably'. The logic of the dominant system of provision is considered decisive for the behaviours of all actors involved. Transforming these systems is considered unlikely, as the division of power and the internal logic (driven by a search for short-term profit) of the conventional market-based food-supply system are considered to contradict the need to increase sustainability. According to this perspective, a more promising strategy towards greening food provision would be to create an alternative system for accessing food through direct exchange, active consumer participation and additional opportunities for political engagement. Local food networks, CSA, etc., which intend to favour 'socially embedded' economies of place, reduce environmental impacts of food provision and build or strengthen local communities (Lebel and Lorek, 2008). Such alternative local food networks constitute a grassroots response to economic globalization and its environmental impacts (see also Chapter 6):

> *These responses are multi-dimensional, and create space for the expression of different sets of values, objectives and motivations than is possible within the conventional economy. As such, they are valuable experimental niches, and they are the repository of some of the more radical transformative impulses for sustainable consumption.* (Seyfang, 2006, p394)

It is difficult, however, to reconnect food consumers and producers because consumers actually value this in the context of their already-routinized food consumption practices (Hinrichs, 2000; Eden et al, 2008). Besides creating alternative supply chains, only interventions by governments to control large private companies can be expected to influence the dynamics of the powerful food-provision system. So, although the first analytical perspective presented here focuses on individual consumers' characteristics, this second view stresses the structural traits of the market system that limit possibilities for changes from within.

Both perspectives, however, show substantial weaknesses, and there is a need to move beyond this individual agency–social system dichotomy. Analysing consumer behaviour from an individual perspective remains inadequate. Building on well-known attitude and behaviour models (Ajzen, 1991) has its limits, despite the increasing attention that researchers are giving to the social context of behaviour, because the correlations between values and behaviours remain weak. Studies of consumer behaviours from a systems perspective are repeatedly confronted with unexpected results because consumers' behaviours are much more diverse that the theoretical model would predict.

## Social-practice perspectives on food consumption

A third perspective for analysing consumers' roles in food provision, the social-practices model, tries to find a balance between the individualist and structuralist approaches considered so far. Human action must be understood as a shared 'praxis' because humans are involved in social practices that they share with others. In social practices, the dynamics are determined by individual and system logic at the same time as the individual and the structure are indissolubly connected and neither has logical primacy over the other. The social-practice model therefore focuses on behaviour as a social activity situated in time and space (Spaargaren, 2003). Social structures are both enabling and restrictive. Individual consumer behaviour is understood in close conjunction with the social structure of provisioning goods and services. Food consumption is an example of routinized social practices, but these practices consist of not only simply buying and using goods and services, but the pursuance of 'social constructs like comfort, convenience, hygiene, nutrition' (Wilk, 2010, p46; see also Shove, 2003). People are consuming to achieve other ends, such as fulfilling social obligations or gender-role expectations (Wilk, 2010); therefore, consumption is more than a goal in itself.

People are involved in various social practices, and the assembly of these social practices, combined with the 'narrative' (the story) that connects them, is called their lifestyle. Lifestyles refer to individual, but also shared, stories that give coherence and significance to the different social practices in which people participate. People are engaged in creating and reproducing meanings in daily life by 'attempting to knit together the different experiences and roles of life' (Halkier, 2001, p802). Empirically, lifestyles combine social status, attitudes and behaviour. Concrete social practices are connected to specific social structures and environmental heuristics, which cluster into different (sustainable) lifestyles. Therefore, *the* environmentally conscious citizen does not exist, and we should link an individual's disposition to environmental change to opportunities in social structures (see Box 11.3).

Food entails a particular domain of consumption because eating is a highly *routinized* social practice. This becomes particularly visible on the shopping

## Box 11.3 Consumers and lifestyles in organic food consumption

In a study to identify the potential for increasing the market share for organic food, a marketing company in Germany studied consumers and their lifestyles. In the past, consumers were mostly classified on the basis on their income, but this study classified consumers based on a combination of income and values, which enabled researchers to identify a highly elaborate differentiation among consumers. By identifying lifestyle groups that may be interested in organic food, it becomes possible to develop much more targeted marketing strategies (see Figure 11.1).

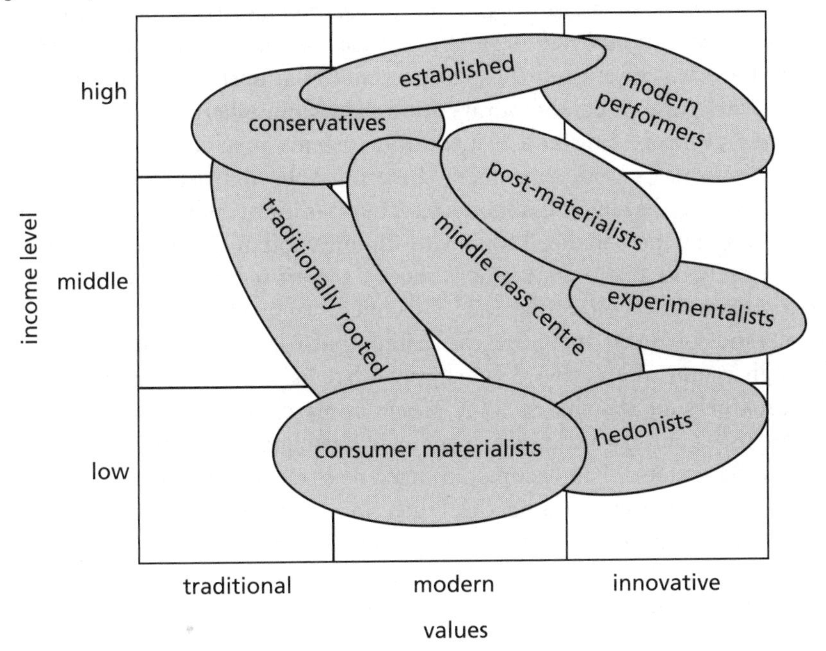


**Figure 11.1** *Lifestyle groups*

*Source:* Based on Bio Verlag (2006)

Researchers found that of all the German lifestyle groups, 'middle class centre', 'modern performers' and 'post-materialists' were the most interested in buying and consuming organic food products.

*Source:* Bio Verlag (2006)

floor of supermarkets, where consumers purchase familiar products and well-known brands. This does not mean, however, that food consumption is fixed and stable forever. Changes in social practices may be triggered by the introduction of alternative meanings of food, unknown conduct, unfamiliar products or reconfigured infrastructures. Consumers' handling of food quality is socially and culturally broader than the cognitive rationality that some researchers assume and includes religious, ethical, social and other considerations. Food consumption necessarily involves an intimate and complex relationship between man and nature, i.e. eating makes food part of the human body, turning food into a particularly sensitive issue (Beardsworth and Keil, 1997). Consuming food requires people to permanently trust its quality and safety because this cannot be presupposed. This trust must be actively maintained and is not simply a given even when it is established (Kjaernes et al, 2007). Trust can be established through face-to-face relationships between the sellers and buyers of food, but also through scientific research, management systems and governmental regulations. In the first case we talk about *personalized* trust and in the second case about *institutionalized* (or *abstract*) trust (Giddens, 1990). Although most trust in food was organized through personal relationships in the past, this in no longer the case in contemporary food provision, where long-distance trade and complex processing and distribution systems preclude direct interaction between producers and consumers. Institutionalized trust, based on expert systems, is necessary in global modernity, but it is more ambiguous and vulnerable to crisis than personalized trust. Consuming food in global modernity forces people to rely on abstract systems, scientific expertise and various information systems when making long-distance assessments of products and on the production methods involved, including the sustainability impacts resulting from them. Although many alternative food-supply systems try to (re)create consumer trust on a personalized basis, some others apply certification in an attempt to secure trust in an institutionalized manner. Ambiguity, or balancing trust and risk in food systems, has become an ingrained trait of the contemporary food supply, making many consumers feel uncertain about whom and what to trust. When interacting with expert systems and their representatives, lay persons such as consumers maintain some reservation or ambivalence, and their trust may only be temporary. Consumer trust in (information about) food evolves through interactions with wider societal dynamics and can acquire various shapes depending on multiple factors. Consumer engagement with the system of food provision displays a 'structural ambivalence' – multiple meanings, dilemmas and negotiations. This stems from the combination of a desire to use food to construct a domestic realm with some autonomy from the public sphere, and their reliance upon market-based systems of provision to realize that desire (Cook et al, 1998).

Even routinized social practices, such as food consumption, may be disrupted through unexpected incidents. A food crisis, but also important biographical changes, may interrupt daily food-consumption routines (de-routinization). New practices then are consciously built in a temporary reflexive phase, wherein communication from authorities and food-chain actors is compared with independent sources, family, friends and colleagues. These new practices, in turn, can develop into new routines (re-routinization) (Spaargaren, 2003). These different practices must be actively translated within the consumer's own social context, biography and everyday life. These changing practices may entail various elements, as consumers may decide to change their eating habits, buy alternative food products within the conventional channels or through alternative systems of provision (direct sale, certified food).

Modern-day food provision makes direct interaction between producer and consumer increasingly elusive because of the multiple phases through which food products must pass before they arrive at the consumers' plate. But, although food-supply chains cover a number of stages, from primary production to final consumption, whereby each can be considered a social practice, the location where the consumer actually meets the system of provision deserves particular attention. It is at this point of access, or the *consumption junction* (Schwartz-Cowan, 1987), where the world of food providers connects with the worlds of consumers. The world of providers is dominated by the logic of (science-based, economically rational) systems and has to be linked with the domestic logic of end-users, the worlds of consumers and their households, with their particular concerns and everyday practices. As shown in Figure 11.2, it is at these points of access where consumers and providers meet, as do their respective logics.

Consumers are human agents characterized by different concerns and lifestyles, while providers are representatives of the system of provision, which involves the stages of food production, processing and trade. Despite the obvious power imbalance between the two sides, this does not mean that consumers are powerless and simply victims of the system of provision. Below we pay particular attention to possible sources of power that consumers can build on when trying to change this system. But providers can also engage in sustainability transitions in different ways (Spaargaren and Van Koppen, 2009). They can make their production processes and products more sustainable, but also actively seek to place their sustainability initiatives and responsibilities in

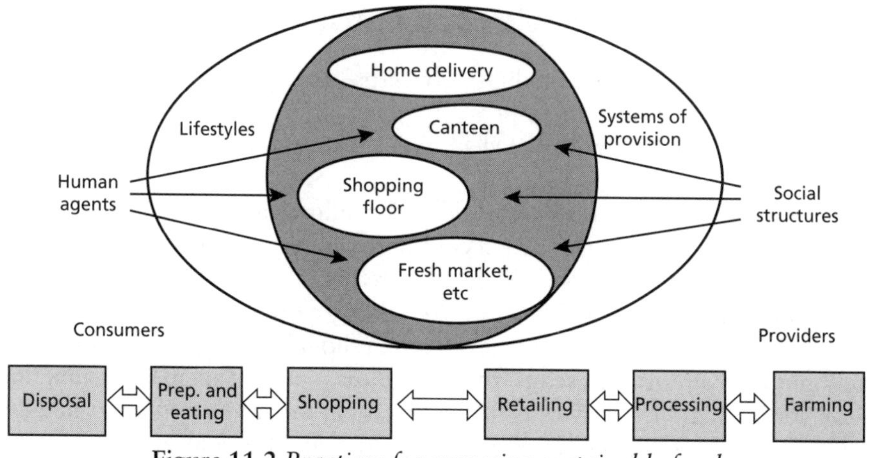


**Figure 11.2** *Practices for accessing sustainable food*

*Source:* The authors and Spaargaren (2003)

## Box 11.4 Definitions of sustainable consumption

Sustainable consumption is:

*The use of goods and services that respond to basic needs and bring a better quality of life, while minimizing the use of natural resources, toxic materials and emissions of waste and pollutants over the lifecycles, so as not to jeopardize the needs of future generations.*

S. Ofsted, Symposium: Sustainable Consumption (1994)

*The special focus of sustainable consumption is on the economic activity of choosing, using, and disposing of goods and services and how this can be changed to bring social and environmental benefit.*

International Institute of Environment and Development (1998)

*Sustainable consumption is not about consuming less, it is about consuming differently, consuming efficiently, and having an improved quality of life.*

United Nations Environment Programme (1999)

*Sustainable consumption is a balancing act. It is about consuming in such a way as to protect the environment, use natural resources wisely and promote quality of life now, while not spoiling the lives of future consumers.*

*National Consumer Council (UK) (2003)*

*Source:* Jackson (2006)

the broader context, promoting more sustainable lifestyles and consumption patterns. The increasing availability of green food in conventional distribution channels is directly connected to the behaviours of proactive providers, who, challenged by governmental regulations and societal pressures, try to create markets and level playing fields for new products and services (see Chapter 10 for a further elaboration of the role of providers, especially retailers).

## Sustainable Food Consumption

Although food-related sustainability concerns among consumers seem to be growing, it remains difficult to precisely define sustainable consumption with respect to food products. Four definitions of sustainable consumption are mentioned in Box 11.4.

The coexistence of these various definitions illustrates the complexity of the debate surrounding sustainable food consumption. An important element in this debate is whether the focus should be on increasing the environmental efficiency of existing consumption patterns through relying more on sustainable products and production processes, or whether attention should also be given to changes in consumption patterns themselves. In other words, does sustainable consumption mean consuming differently or consuming less? In order to avoid intervening in the private domain of consumers and their individual freedom to choose, many connect sustainable consumption with increasing the efficiency of how natural resources are used rather than with reducing volumes of consumption.[5]

## Box 11.5 Multiple consumer concerns about the sustainability of food

Consumer concerns related to the sustainability of food mostly consist of varying (combinations of) the following:

- naturalness – unadulterated and 'natural' food;
- food safety – safe from conventional and modern food risks;
- animal welfare – respecting the rights/integrity of animals;
- environmental (ecosystem) – respecting present ecosystems and the rights of future generations;
- social impacts on workers and (small) farmers producing food.

Not all consumers share all of these sustainability concerns, but most of them do at least to a certain extent at some time in their lives, with large variations among different regions and different categories of consumers.

*Source:* Oosterveer et al (2007)

Consumers therefore necessarily play a crucial role in creating more sustainable food-consumption practices. Consumers have always been concerned about the safety, quality, taste and price of their food. The industrialization of food processing has increased such worries, and the addition of chemical substances to many food products since the 1970s and the introduction of GMOs in the 1990s have generated even more public debates on the dangers involved in the food-production process and the reliability of official regulations. These concerns have been recently supplemented by environment-related topics that vary and evolve over time, such as climate change, biodiversity and animal welfare. Hence, nowadays, more and more consumers define food quality beyond simply the objective characteristics of food products and include the (in)direct environmental, animal welfare and social impacts as well. Environment-related concerns have supplemented and adapted consumer worries about the safety, quality, taste and price of food. Environment-related food concerns are hard to define in general terms because they vary and evolve over time, but it is possible to distinguish five motivational dimensions that are relevant for most consumers and not mutually exclusive (see Box 11.5 for these consumer concerns).

Consumers' expressions of sustainability concerns and environmental changes within the system of provision must be linked at the point of access (see Figure 10.2) in order to (re)construct sustainable food-consumption practices. Such practices involve selling and buying sustainable products; sharing information about production, processing and trade; and communicating about sustainability concerns, sustainability narratives and methods of use. Consumers' greatest concerns vary considerably, affected by time, place, life-cycle stages and other factors. This multi-dimensionality of green-food consumerism needs to be acknowledged, but it creates challenges. Different concerns can be satisfied in distinctive ways, generating the need to compare them and encouraging more conscious decision-making. Moreover, as consumers make different choices with regard to their individual sustainability concerns, the need to justify these concerns increases as well.

Various combinations of practices, behaviours and narratives with regard to sustainable food can be categorized as representing different green lifestyles. Therefore, the recent trend towards a greater diversity in lifestyles has a considerable impact on food-consumption practices as well. For each food-consumption practice and category of food products, a range of specific consumer concerns can be formulated, which providers can take into account when developing their provisioning strategies. Consumers may also differ in the (relative importance of the different dimensions of) sustainability concerns expressed in different settings. They may express other concerns in the family context at the dinner table than they do in the professional context of the company canteen, but also concerning different kinds of food products, such as those related to fresh fruits and vegetables or to ready-made meals. As a result, understanding the importance of sustainability in different consumer practices is complex and requires further analyses of activities at the locations where consumers access food. Consumers may buy organic food in conventional supermarkets, but they may also actively engage in creating direct supply chains, or even take up some of the producer tasks themselves, such as harvesting or transporting food. Consumers may also combine buying some food products in specialized sustainable retail shops and other food products in regular supermarkets. This makes some observers complain about the lack of consistency in consumer behaviour or worry about the 'unmanageable consumer' (Gabriel and Lang, 1995). Researchers can analyse consumer behaviours as embedded in wider social practices, rather than in isolation, to help reduce this confusion.

Greening food consumption thus is not simply a matter of voluntarism, of individual (consumer) choice. Analysing sustainable food consumption applying the social-practices perspective means studying a complex set of behaviours carried out in various contexts by people from different social and motivational backgrounds. As a result, different social practices may emerge in buying (sustainable) food related to environmental concerns, lifestyles and lifestyle politics. If we look at only buying sustainable food, we can already distinguish four different channels: direct sales via farmers' markets and shops, home delivery and drop-off systems, the specialty shops of the alternative circuit and the conventional retailing models of supermarkets, restaurants and food services (canteens, retailing, etc.). Different (sustainable) food products are on offer in these channels, and within them different narratives on sustainable food have developed. Each of these channels constitutes an arrangement of social actors, food products and (narratives related to) sustainability concerns.[6] The presence of a large number of access points and multiple sustainability-related consumer concerns contribute to making food-consumption practices more reflexive than in the past.

## Promoting Sustainable Food Consumption

Innovative governance arrangements to promote sustainable food consumption emerge next to and beyond national governmental regulations. Widening consumer concerns, combined with an increasing dependency on global dynamics, creates particular challenges for conventional nation state-based regulatory practices. Nation states seem unable to incorporate these concerns because of their limited capacities to intervene beyond their national borders; and thus more consumers are becoming actively engaged in developing alternative governance arrangements. Together with civil-society organizations, consumers become co-drivers of creating sustainable food-consumption practices. Consumers can become involved as ecological citizens, as political consumers and through life(style) politics (Spaargaren and Mol, 2008). As political consumerism is a recent conceptual tool to analyse consumer roles in global modernity, this deserves particular attention and is addressed in the following section.[7] An alternative, more top-down approach to promoting agri-food sustainability emphasizes eco-labelling, especially with respect to 'organic', for food products.

### Political consumerism

The growing consumer involvement in food-governance arrangements shows that citizenship and consumerism can no longer be considered as only opposing practices and discourses. Today, no clear boundaries can be drawn around 'public' citizenship and 'private' consumerism. In their consumer role, critical citizens practice new forms of action that are outside the traditional formal political arena but within the 'sub-political' sphere (Beck, 1992). This political consumerism (Micheletti, 2003) refers to voluntary, ad-hoc-organized, mostly civil society-based forms of (environmental) political action. The concept brings together the different forms of politics that connect environmental activities from upstream actors of production–consumption chains and networks more directly and visibly with the interests and activities of citizen-consumers at the lower end of these chains and networks, and vice versa. Consumers can become co-drivers of global environmental change by using their power through diverse networks that comprise actors and interests from civil society, the market and the state.[8] These consumers are not rejecting the consumer role in order to replace it with a citizen role, but they are connecting the politics of consumption with the practices of being a responsible consumer (Barnett et al, 2011). Sustainable consumption seeks to embed social and environmental commitments into the rhythms and routines of everyday life (aligning abstract principles with everyday practices), but it is an organized field of strategic interventions at the same time. The growth of political consumerism is the outcome of organized efforts by a variety of collective actors to rearticulate the ordinary ethical dispositions of everyday consumption. Political consumerism has two dimensions: the process of *discursive* engagement with the frames of reference that already shape people's consumer behaviours and the process of using various *devices* to enable people to readjust their consumption behaviour. Campaigns for sustainable consumption are not oriented to completely changing people's identity into that of 'ethical consumers', but at the generation of various singular actions. Sustainable consumption is thus not an individual or moral act, but an organized field of strategic mobilization and campaigning. It seeks to embed people's existing dispositions of care, concern and solidarity into the global politics of mobilization, activism, lobbying and campaigning around issues of sustainability. It is 'a means to an end rather than the end in itself' (Barnett et al, 2011, p201).

Critical food consumers may use their shopping-bag power to make conscious choices in their everyday lives, thereby directly addressing the market's organization and combining public and private aspects of altruism. A particular category of food products is offered to those consumers through the inclusion of specific meanings (regarding ecology, solidarity, fairness, and so on) that go beyond the objectively verifiable characteristics of the product itself. By using labels and standards, new connections can be made between food producers and consumers that surpass the economic relationship of buyer and seller.[9] For some, such food labels are simply a market tool intended to offer consumers a choice and allow them to make buying decisions according to how they want to live their lives. But others understand food labels as an instrument that can address wider concerns and co-create alternative economies. Consumer engagement in governing the global food supply is therefore not necessarily based only on product prices or 'objective' information about the composition of particular food products, but may also comprise more intangible values, such as 'sustainability', 'social equity' and 'quality', which are facilitated by the availability of tools such as labels. Exercising this role of the political consumer may be demanding in terms of knowledge and information and in shopping practices (Young et al, 2009), and therefore, not all consumers are willing to engage in this process every time they shop. One response is therefore to select a broad label that simplifies this demanding condition, and one example of this is consuming organic food.

## Organic food consumption

The introduction and expansion of organic food constitutes an interesting case for analysing sustainability in food consumption. In 30 years, organic food has evolved from a minor alternative method for producing food to the dominant reference model for consuming 'natural and sustainable' food. Organic food seems to encapsulate a response to several consumer concerns, particularly those related to 'naturalness'. Over the years, it has become the preferred response to consumer worries about the ongoing industrialization of the dominant food-provisioning system, with its use of chemicals and processing techniques. Organic food offers an alternative to the dangers of pesticide use, factory farming of animals, industrial processing of foodstuffs, degradation of rural landscapes and food safety risks (BSE, GM, etc.), which are all considered as characteristics of the conventional food system. Presenting a concrete alternative, organic food encourages consumers to reflect on their buying and eating habits because they can also choose differently.

Organic farming, whose principles originated in Europe in the 1920s and 1930s, became popularized in the 1970s, 'criticizing the destructive nature of agro-industrial practices and creating local production/distribution/consumption systems linking small-scale organic farms, distribution via food cooperatives, box schemes, and farmers markets, and wholesome diets' (Raynolds, 2004, p729). Since the 2000s, organic farming has spread to other parts of the world, although the main consumer markets have remained in the more advanced economies, despite some interest from the growing urban middle classes in non-OECD countries.

Especially during its pioneering stage, alternative relationships between producers and consumers were attempted within the organic sector. Initiatives, such as box schemes and CSA, played an important role in the organic food supply's development. Organic food-subscription schemes (box schemes) were developed, primarily in Europe, so that consumers could gain access to organic food and to guarantee organic farmers a reliable market. Every week, (mainly) fresh organic produce is packed into a box and delivered to a place where subscribing consumers can collect it. CSAs are mainly a US phenomenon, although they are also present in other countries. The purpose of CSAs is to build a community of people supporting a farm operation, whereby growers and consumers provide mutual support and share the risks and benefits of food production. Typically, the members cover the costs of the farm operation in advance and in return receive the farm's produce. In addition, these supporters can reconnect with the farmland and even participate in food production themselves (DeMuth, 1993).

In recent decades, these initiatives have become more consolidated and institutionalized, via the founding of IFOAM, the formulation of national standards and certification systems for organic agriculture (Boström and Klintman, 2008) and growing sales through conventional retail operations, including supermarkets. Nevertheless, despite its popularity, organic food still represents only a small percentage of food sales.

Smith (2006), who analysed organic food's historical trajectory in the UK, an example of radical change in food provision, distinguishes three phases in the development of the organic niche, starting with a nurturing phase (1946–1970), when activists were developing and promoting organic food. This was followed by a pioneering phase (1970–1990), when organic producers were creating a small organic market, and finally a growth phase (1990–present), when mainstream actors become involved. Similar phases can be observed in other countries, although the exact timing varies.

In these last decades, as the market for organic food expanded, its consumers changed. For instance, in the US some years ago, organic food consumers were

**Table 11.1** *Trends in organic food sales in selected countries (million €)*

| | *1997* | *2003* | *2005* | *2007* |
|---|---|---|---|---|
| Germany | 1600 | 3100 | 3900 | 5300 |
| UK | 445 | 1607 | 2332 | 2560 |
| France | 508 | 1578 | 2200 | 1900 |
| Netherlands | 167 | 395 | 420 | 500 |
| US | 2662 | 7690 | 10,245 | 14,000 |

*Source:* Willer and Yussefi (2004); Willer et al (2008); Biologica (2008)

typically from well-educated, health- and environmentally concerned, white, middle-class families (Dimitri and Oberholtzer, 2006). Nowadays, they make up a much broader group with different income levels, ethnic backgrounds and age ranges. Over time, the environment seems to have declined as the motivating factor for buying organic food, while health, taste and food safety have become the more dominant reasons. Empirical research shows that average consumer knowledge about the definition of organic food and its formal requirements is limited and that consumers often use 'organic' as a proxy for 'natural' food (Aarset et al, 2004).[10] This shift occurred in combination with the rapid increase in organic food sales in the last 10 to 15 years. Table 11.1 shows the substantial growth in organic food sales in selected countries in recent years, which is remarkable in the context of the saturated food market that is characteristic of most developed economies.

The growth in sales therefore also translates into an increasing market share (see Table 11.2), which in most cases still does not exceed 5 per cent (except for Austria and Denmark, with 5.3 and 6.0 per cent respectively).[11]

Consumers are willing to pay premium prices for organic food because they consider it to have been produced in a sustainable, environmentally sound manner and that buying also means supporting family farms. Actual spending for organic food per consumer is growing, but remains small compared to what consumers combined spend on food in general (see Table 11.3 for an overview of the annual expenses for organic food).

**Table 11.2** *Changes in market share for organic food in selected countries (%)*

| | *1997* | *2003* | *2005* | *2008* |
|---|---|---|---|---|
| Germany | 1.2 | 2.2 | 3.0 | 3.4 |
| UK | 0.4 | 1.7 | 2.3 | na |
| France | 0.4 | na | 1.1 | 1.7 |
| Netherlands | na | 1.0 | 1.8 | 2.1 |
| US | 0.8 | 1.9 | 2.5 | 3.5 |

*Note:* na = no data available.

*Source:* Lohr (2001); www.organic-world.net (accessed 8 October 2009)

**Table 11.3** *Countries with the highest per capita consumption of organic food (most 2008)*

| *Rank* | *Country* | *Annual per capita consumption of organic food (in €)* |
|---|---|---|
| 1 | Denmark | 132 |
| 2 | Switzerland | 119 |
| 3 | Austria | 97 |
| 4 | Luxemburg | 85 |
| 5 | Liechtenstein | 85 |
| 6 | Germany | 71 |
| 7 | Sweden | 68 |
| 8 | US | 49 |
| 9 | Canada | 41 |
| 10 | UK | 41 |
| 11 | France | 41 |
| 12 | New Zealand | 33 |
| 13 | Italy | 33 |
| 14 | Netherlands | 33 |

*Source:* Willer and Kilcher (2010)

Over time, increasing interaction has taken place between the organic and conventional food systems, maintaining the basic principles of organic production but rising to the convenience, availability and cost demands of conventional practices. Buying organic food in supermarkets clearly fits into this trend towards convenience. As Table 11.4 indicates for the case of Europe, the relative market shares of supermarkets in organic food sales range widely, but there is an overall shift towards buying through supermarkets in all countries except for Italy. At the same time, several European countries, such as Germany, France, Italy and the Netherlands, still have a considerable number of specialized organic retailing stores.

Selling organic food via supermarkets and increasing the scale of organic-food production have generated heated debates among scientists and activists. A part of the organic-food movement accepted that its provision was mainstreamed, while another part insisted on the need for close connections

**Table 11.4** *Market shares for organic food by distribution channel in selected European countries (%)*

| *Country* | *Supermarkets* | | *Specialty stores* | | *Direct sales* | |
|---|---|---|---|---|---|---|
| | *1999* | *2008* | *1999* | *2008* | *1999* | *2008* |
| France | 45 | 39 | 45 | 37 | 10 | 24 |
| Italy | 25–33 | 25 | 33 | 52 | 33–42 | 20 |
| Germany | 25 | 49 | 45 | 28 | 20 | 23 |
| Netherlands | 20 | 43 | 75 | 43 | 5 | 14 |
| Switzerland | 60 | 76 | 30 | 15 | 10 | 9 |
| UK | 65 | 75 | 17.5 | 13 | 17.5 | 12 |

*Source:* Lohr (2001); Richter (2009)

### Box 11.6 Oxfam's diet to benefit people and planet

In a briefing paper, the development organization Oxfam invites consumers in wealthier countries to consider the social and environmental impacts of their food choices. Producing food provides a vital source of income for millions of poor farmers and workers, but producing, processing and trading food all have negative impacts on climate change. The NGO therefore suggests consumers take four simple actions every week:

- waste less food;
- reduce consumption of meat and dairy products;
- buy fair trade produce;
- buy other foods from developing countries.

The first three actions seem self-evident. With respect to the fourth, Oxfam argues that agro-exports from developing countries improve poor people's income and food security, while the distance food travels is a poor measure of its total environmental impact.

*Source:* Oxfam GB (2009)

between organic food producers and consumers via alternative supply networks to maintain an integrated organic vision. Promoters of 'genuine' organic food consider just applying organic production methods as lacking because it is too technocratic and not political enough (Goodman and Goodman, 2001).[12] They insist on the need to maintain local trade and direct social interaction between producers and consumers and to not rely on organic food sold via supermarkets. Others are less concerned about the physical or social distance from producer to consumer, and more concerned about the (lack of) transparency/trust/information exchange between them. 'Instead of having difficulties with how food production is distanced from consumers, we have a problem with how the intermediary assurance organizations are distanced from consumers' (Eden et al, 2008, p1054). The success of organic food may threaten the sector because there are increasing dangers that less motivated producers try to benefit from the attractive profit margins by minimizing or even ignoring (part of) official regulations. Nevertheless, while preferences for local organic food persist, global sales of organic food via supermarkets are growing fast (Raynolds, 2004). In much the same way as they dominate the conventional food retail sector, supermarkets nowadays have also become the main outlet for organic foods (Banks and Marsden, 2001).

Although many consider these two forms of organic food provision (i.e. mainstreamed, minimalist, large scale and distributed via supermarkets versus alternative, orthodox, small scale and sold through direct distribution mechanisms) as necessarily antagonistic, others expect them to reinforce each other. The presence of a small-scale, direct system of organic food supply as an available alternative compels the more conventional system to make sure that it does not significantly undermine the organic food principles. The existence of a supermarket-based distribution system forces direct-sales systems to critically monitor their prices and logistics.

## Sustainable Diets

Another strategy to increase food sustainability is the application of healthy, low-impact diets. Most research confirms that diets composed of primarily fresh fruits and vegetables, particularly when they are of local or regional origin, and with less meat than conventional Western diets, are healthier and have lower environmental impacts than conventional diets (Duchin, 2005). Contemporary diets in wealthy countries differ substantially from such sustainable diets in both quantity of food intake and mix of products. High meat consumption, in particular, is under criticism because it is related to high use of water,[13] while its energy efficiency is very low.[14] Oxfam, the transnational NGO headquartered in the UK, promotes its own diet (see Box 11.6).

Changing diets is not a simple matter, however, because food choices are embedded in household practices, lifestyles and local or national cultures. For instance, the advice to reduce meat consumption is inciting much debate in Europe and the US because many people consider it as being essential for a 'good meal'. Consuming meat is not simply a matter of dietary intake; for many people, it is an important aspect of their identity and not easily changeable. And, as information that matches one's identity is more persuasive than information that is not matched (Abrahamse et al, 2009), it is clearly difficult to convince meat-eaters about the advantages of vegetarian diets.

Government involvement through increasing the price of unsustainable food products via taxes may contribute to changing consumer practices (Duffey et al, 2010). Next to information and awareness-raising, using the price mechanism may be an attractive additional instrument.

## Conclusion

For a long time, consumers have been seen as passive stakeholders without real influence in food provision. But, as the examples of organic food and dietary change make clear, consumers have considerable power to play a key role in steering food provision towards more sustainability. Consumers cannot be ignored, particularly under the conditions of enhancing globalization and proliferating societal concerns about the impacts of food production, processing and trade. Understanding consumer roles requires a social-practices approach, as consumer behaviour is embedded in wider social structures involving other social actors, balances of power, routines and trust. Contemporary food supply shifts the conventional roles of actors, such as governments, producers and retailers, and brings in unfamiliar challenges about food safety and environmental and social impacts. Consumers are responding to these challenges in different ways. While some consumers may attempt to maintain their familiar consumption practices, increasing numbers are looking for innovative responses and buy different food products, experiment with ways to narrow their distance from producers or change their diets.

Different strategies lead to changes in food-consumption practices, some more or less radical, and more or less visible, but they all deserve more in-depth study. Understanding these changes is essential when analysing transformations in the contemporary food supply. Consuming fair trade or sustainably labelled food may be considered 'caring at a distance'. Such arrangements based on inserting social or environmental considerations into food-supply chains are created 'by a whole host of actors in the process of governing consumption and trade' (Hughes et al, 2008, p350). These actors include NGOs, community groups, social-movement organizations and media, which are all in some way related to food advocacy. Market actors today are quick to respond to evolving consumer tastes and concerns, even when overlooked by activists. In an era of 'less government is better government', the role of governmental agencies in encouraging sustainable food-consumption practices continues to evolve, and they increasingly rely on indirect policy instruments, including labelling, certification and educational approaches.

---

### *Take-home lessons*

- Consumer behaviour can be better understood when it is analysed as embedded in social practices and not only as individual choices when buying food.
- Sustainability is becoming an aspect of consumer behaviour, but it is multi-dimensional and shows large variations in content and relative importance.
- Political consumerism offers an interesting innovative strategy in promoting more sustainable food consumption.

---

## Notes

1 Consumption had attracted attention before, as even by the 1960s and 1970s scientists and activists were concerned about consumption as an addiction (e.g. Marcuse, 1964) and over-consumption (e.g. Moore Lappé, 1971). Nevertheless, it would take another two decades before their calls would become incorporated in environmental policies.

2 Between 1920 and 1960, a uniform national eating pattern appeared replacing the many regional differences. After this standard had spread throughout the Netherlands, a process of re-differentiation started again through the increased use of ready-to-use food products made possible by continued income rises and the spread of supermarkets (Van Otterloo, 2000).

3 Variations in cooking skills need more attention because they influence dietary choices (Caraher et al, 1999), and diet matters for environmental sustainability (Goodland, 1997; Duchin, 2005).

4 For instance, in a report by the Australian Food and Grocery Council (AFGC, 2010), 84 per cent of respondents said they were concerned about the impacts of

their purchasing decisions on the world, but only 13 per cent actually bought green.

5 Some scholars, however, consider it rather optimistic to expect increased efficiency in natural resource use to assure sustainable consumption (York et al, 2003). For instance, the gains from more efficient farming methods may easily disappear when consumers in up-and-coming economies such as China adopt more Western diets. Hence, it may be necessary to include more radical changes in lifestyles and consumption patterns if sustainable consumption is really to be achieved.

6 Supermarkets nowadays offer thousands of different food products, including organic, fair trade or otherwise sustainably labelled food products. They may nevertheless differ in their assortment, and with regard to the positioning of sustainable food in their floor plan, their presentation on the shelves, the type of information presented, the kind of images applied and the narratives expressed.

7 For more information on environmental citizenship, see Dobson and Bell (2006). See also Chapter 4 on regulation.

8 It seems that civil-society organizations are currently more capable of establishing trust relationships with consumers than state- and market-based organizations (Oosterveer and Spaargaren, 2011).

9 Some consider that there are too many food labels, while the information they contain may not always be clear. Consumers are, however, unlikely to be interested in all labels. Their attention is selective for the particular issues they are interested in and ignore the broad mass of material (Ipsos Mori, 2010).

10 Smith (2006) suggests that the growth in organic food has been fuelled more by problems in the dominant food production regime than by characteristics of the organic niche itself.

11 It is important to note that market shares for particular products may extend well beyond these overall market shares, e.g. in Denmark organic milk has a market share of more than 20 per cent (Biologica, 2008), and in Germany organic baby foods have a market share of 64 per cent (Bakker and Bunte, 2009).

12 For instance, that 'green' labelling and certification schemes are defined by NGOs and companies in the global North, thereby reinforcing already-existing global power imbalances (Goodman and Goodman, 2001).

13 Producing meat for an average non-vegetarian diet requires 1000 litres of water more per week than is required for producing food for a vegetarian diet (Marlow et al, 2009).

14 The energy efficiency of meat production, calculated as the total energy output in available meat for consumption compared with the total energy input, is estimated to be less than 10 per cent for all meat combined (Eshel and Martin, 2006). For further information on the environmental impact of meat production, see FAO (2006).

## Further Reading

Barnett, C., Cloke, P., Clarke, N. and Malpass, A. (2011) *Globalizing Responsibility: The Political Rationalities of Ethical Consumption*, Wiley-Blackwell, Chichester: an in-depth analysis of ethical consumption (or political consumption, as it is understood in this chapter).

Beardsworth, A. and Keil, T. (1997) *Sociology on the Menu: An Invitation to the Study of Food and Society*, Routledge, London and New York: introduces food consumption from a sociological/anthropological perspective.

Jackson, T. (ed) (2006) *The Earthscan Reader in Sustainable Consumption*, Earthscan, London: an overview of sustainable consumption.

Sassatelli, R. (2007) *Consumer Culture: History, Theory and Politics*, Sage, London: an insightful and theoretically well-grounded presentation of consumers and their social lives.

## References


Aarset, B., Beckmann, S., Bigne, E., Beveridge, M., Bjorndal, T., Bunting, J., McDonagh, P., Mariojouls, C., Muir, J., Prothero, A., Reisch, L., Smith, A., Tveteras, R. and Young, J. (2004) 'The European consumers' understanding and perceptions of the "organic" food regime: The case of aquaculture', *British Food Journal*, vol 106, no 2, pp93–105

Abrahamse, W., Gatersleben, B. and Uzzell, D. (2009) *Encouraging Sustainable Food Consumption: The Role of (Threatened) Indentity*, University of Surrey, Guildford

AFGC (Australian Food and Grocery Council) (2010) *Green Shopper Summary Report 2010*, Net Balance, Melbourne

Ajzen, I. (1991) 'The theory of planned behavior', *Organizational Behavior and Human Decision Processes*, vol 50, no 2, pp179–211

Atkins, P. and Bowler, I. (2001) *Food in Society: Economy, Culture, Geography*, Arnold, London

Bakker, J. and Bunte, F. (2009) *Biologische internationale handel*, LEI, The Hague

Baltussen, W., Wertheim-Heck, S., Bunte, F., Tacken, G., Van Galen, M., Bakker, J. and Winter, M. (2006) *Een Biologisch Prijsexperiment; Grenzen in zicht?*, LEI, The Hague

Banks, J. and Marsden, T. (2001) 'The nature of rural development: The organic potential', *Journal of Environmental Policy & Planning*, vol 3, pp103–121

Barnett, C., Cloke, P., Clarke, N. and Malpass, A. (2011) *Globalizing Responsibility: The Political Rationalities of Ethical Consumption*, Wiley-Blackwell, Chichester

Beardsworth, A. and Keil, T. (1997) *Sociology on the Menu: An Invitation to the Study of Food and Society*, Routledge, London and New York

Beck, U. (1992) *Risk Society: Towards a New Modernity*, Sage, London

Biologica (2008) *Bio-Monitor: Jaarrapport '08*, Biologica, Utrecht

Bio Verlag (2006) *BioHandel*, Bio Verlag, Aschaffenburg

Boström, M. and Klintman, M. (2008) *Eco-standards, Product Labelling and Green Consumerism*, Palgrave MacMillan, Houndmills

Caraher, M., Dixon, P., Lang, T. and Car-Hill, R. (1999) 'The state of cooking in England: The relationship of cooking skills to food choice', *British Food Journal*, vol 101, no 8, pp590–609

Center for Sustainable Systems (2009) *US Food System Factsheets*, University of Michigan, Ann Arbor

Cheng, S., Olsen, W., Southerton, D. and Warde, A. (2007) 'The changing practice of eating: Evidence from UK time diaries, 1975 and 2000', *The British Journal of Sociology*, vol 58, no 1, pp39–61

Cook, I., Crang, P. and Thorpe, M. (1998) 'Biographies and geographies: Consumer understandings of the origins of foods', *British Food Journal*, vol 100, no 3, pp162–167

Counihan, C. and Esterik, P. (eds) (1997) *Food and Culture: A Reader*, Routledge, New York and London

DeMuth, S. (1993) *Community Supported Agriculture (CSA): An Annotated Bibliography and Resource Guide*, USDA, Washington, DC

Dimitri, C. and Oberholtzer, L. (2006) 'A brief retrospective on the US organic sector: 1997 and 2003', *Crop Management*, www.plantmanagementnetwork.org/pub/cm/symposium/organics/Dimitri (accessed 22 March 2011)

Dobson, A. and Bell, D. (eds) (2006) *Environmental Citizenship*, The MIT Press, Cambridge, MA

Duchin, F. (2005) 'Sustainable consumption of food: A framework for analyzing scenarios about changes in diets', *Journal of Industrial Ecology*, vol 9, no 1/2, pp99–114

Duffey, K., Gordon-Larsen, P., Shikany, J., Guilkey, D., Jacobs, D. and Popkin, B. (2010) 'Food price and diet and health outcomes', *Archives of Internal Medicine*, vol 170, no 5, pp420–426

Eden, S., Bear, C. and Walker, G. (2008) 'Mucky carrots and other proxies: Problematising the knowledge-fix for sustainable and ethical consumption', *Geoforum*, vol 39, pp1044–1057

Eshel, G. and Martin, P. (2006) 'Diet, energy, and global warming', *Earth Interactions*, vol 10, no 9, pp1–17

FAO (Food and Agriculture Organization of the United Nations) (2006) *Livestock's Long Shadow; Environmental Issues and Options*, FAO, Rome

Fine, B., Heasman, M. and Wright, J. (1996) *Consumption in the Age of Affluence: The World of Food*, Routledge, London and New York

Gabriel, Y. and Lang, T. (1995) *The Unmanageable Consumer: Contemporary Consumption and its Fragmentation*, Sage, London

Giddens, A. (1990) *The Consequences of Modernity*, Stanford University Press, Stanford

Goodland, R. (1997) 'Environmental sustainability in agriculture: Diet matters', *Ecological Economics*, vol 23, pp189–200

Goodman, D. and Goodman, M. (2001) 'Sustaining foods: Organic consumption and the socio-ecological imaginary', in M. Cohen and J. Murphy (eds) *Exploring Sustainable Consumption: Environmental Policy and the Social Sciences*, Pergamon Elsevier, Amsterdam, pp97–119

Halkier, B. (2001) 'Risk and food: Environmental concerns and consumer practices', *International Journal of Food Science and Technology*, vol 36, pp801–812

Hinrichs, C. (2000) 'Embeddedness and local food systems: Notes on two types of direct agricultural market', *Journal of Rural Studies*, vol 16, pp195–303

Hobson, K. (2002) 'Competing discourses of sustainable consumption: Does the 'rationalisation of lifestyles' make sense?', *Environmental Politics*, vol 11, no 2, pp95–120

Howes, D. (ed) (1996) *Cross-Cultural Consumption: Global Markets, Local Realities*, Routledge, London and New York

Hughes, A., Wrigley, N. and Buttle, M. (2008) 'Global production networks, ethical campaigning, and the embeddedness of responsible governance', *Journal of Economic Geography*, vol 8, pp345–367

Ipsos Mori (2010) *Qualitative Research to Explore Peoples' Use of Food Labelling Information*, Food Standards Agency, London

Jackson, T. (2005) *Motivating Sustainable Consumption: A Review of Evidence on*

*Consumer Behaviour and Behavioural Change*, Centre for Environmental Strategy, Surrey

Jackson, T. (ed) (2006) *The Earthscan Reader in Sustainable Consumption*, Earthscan, London

James, A. (1996) 'Cooking the books: Global or local identities in contemporary British food cultures?', in D. Howes (ed) *Cross-cultural Consumption: Global Markets, Local Realities*, Routledge, London and New York, pp77–92

Kjaernes, U., Harvey, M. and Warde, A. (2007) *Trust in Food: A Comparative and Institutional Analysis*, Palgrave MacMillan, Houndmills

Lebel, L. and Lorek, S. (2008) 'Enabling sustainable production-consumption systems', *Annual Review of Environment and Resources*, vol 33, pp241–275

Lohr, L. (2001) 'Factors affecting international demand and trade in organic food products', in A. Regmi (ed) *Changing Structure of Global Food Consumption and Trade*, ERS/USDA, Washington, pp67–79

Lunt, P. (1995) 'Psychological approaches to consumption: Varieties of research – past, present and future', in D. Miller (ed) *Acknowledging Consumption: A Review of New Studies*, Routledge, London and New York, pp238–263

Marcuse, H. (1964) *One-dimensional Man: Studies in Ideology of Advanced Industrial Society*, Routledge, London

Marlow, H., Hayes, W., Soret, S., Carter, R., Schwab, E. and Sabaté, J. (2009) 'Diet and the environment: Does what you eat matter?', *American Journal of Clinical Nutrition*, vol 89, pp1699S–1703S

Mestdag, I. (2005) 'Disappearance of the traditional meal: Temporal, social and spatial destructuration', *Appetite*, vol 45, pp62–74

Micheletti, M. (2003) *Political Virtue and Shopping: Individuals, Consumerism, and Collective Action*, Palgrave MacMillan, New York

Miller, D. (ed) (1995) *Acknowledging Consumption: A Review of New Studies*, Routledge, London and New York

Millstone, E. and Lang, T. (2003) *The Atlas of Food: Who Eats What, Where and Why*, Earthscan, London

Moore Lappé, F. (1971) *Diet for a Small Planet*, Ballantine Books, New York

Oosterveer, P. and Spaargaren, G. (2011) 'Organising consumer involvement in the greening of global food flows: The role of environmental NGOs in the case of marine fish', *Environmental Politics*, vol 20, no 1, pp97–114

Oosterveer, P., Guivant, J. and Spaargaren, G. (2007) 'Shopping for green food in globalizing supermarkets: Sustainability at the consumption junction', in J. Pretty, A. Ball, T. Benton, J. Guivant, D. Lee, D. Orr, M. Pfeffer and H. Ward (eds) *The Handbook of Environment and Society*, Sage, London, pp411–428

Oxfam GB (2009) *4-a-Week: Changing Food Consumption in the UK to Benefit People and Planet*, Oxfam GB, Oxford

Padel, S. and Foster, C. (2005) 'Exploring the gap between attitudes and behaviour: Understanding why consumers buy or do not buy organic food', *British Food Journal*, vol 107, no 8, pp606–625

Pollan, M. (2008) *In Defence of Food: The Myth of Nutrition and the Pleasures of Eating*, Allen Lane/Penguin, London/New York

Raynolds, L. (2004) 'The globalization of organic agro-food networks', *World Development*, vol 32, no 5, pp725–743

Roberts, P. (2009) *The End of Food: The Coming Crisis in the World Food Industry*, Bloomsbury, London

Richter, T. (2009) 'Trends in organic retailing 2008', presentation at BioFach Kongress 2009, Nürnberg Messe Convention Centre, Nuremberg, Germany, 19–22 February, http://orgprints.org/15482/2/richter-2009-trends.pdf (accessed 3 February 2011)
Sassatelli, R. (2007) *Consumer Culture: History, Theory and Politics*, Sage, London
Schwartz-Cowan, R. (1987) 'The consumption junction: A proposal for research strategies on the sociology of technology', in W. Bijker, T. Hughes and T. Pinch (eds) *The Social Construction of Technological Systems: New Directions in the Sociology and History of Technology*, The Guilford Press, London, pp261–280
Seyfang, G. (2006) 'Ecological citizenship and sustainable consumption: Examining local organic food networks', *Journal of Rural Studies*, vol 22, pp383–395
Shepherd, R., Magnusson, M. and Sjödén, P. (2005) 'Determinants of consumer behaviour related to organic foods', *AMBIO*, vol 34, no 4/5, pp352–359
Shove, E. (2003) *Comfort, Cleanliness and Convenience: The Social Organization of Normality*, Berg, Oxford
Shove, E., Trentmann, F. and Wilk, R. (eds) (2009) *Time, Consumption and Everyday Life: Practice, Materiality and Culture*, Berg, Oxford and New York
Smith, A. (2006) 'Green niches in sustainable development: The case of organic food in the United Kingdom', *Environment and Planning C*, vol 24, pp439–458
Spaargaren, G. (2003) 'Sustainable consumption: A theoretical and environmental policy perspective', *Society and Natural Resources*, vol 16, pp687–701
Spaargaren, G. and Mol, A. (2008) 'Greening global consumption: Redefining politics and authority', *Global Environmental Change*, vol 18, no 3, pp350–359
Spaargaren, G. and Van Koppen, C. (2009) 'Provider strategies and the greening of consumption practices: Exploring the role of companies in sustainable consumption', in H. Lange and L. Meier (eds) *The New Middle Classes: Globalizing Lifestyles, Consumerism and Environmental Concern*, Springer, Dordrecht, pp81–100
Tansey, G. and Worsley, T. (1995) *The Food System: A Guide*, Earthscan, London
Van Otterloo, A. (2000) 'Voeding in verandering', in J. Schot, H. Lintsen, A. Rip and A. Albert de la Bruhèze (eds) *Techniek in Nederland in de Twintigste eeuw; landbouw, voeding*, Stichting Historie der Techniek and Walburg Pers, Zutphen, pp237–247
Vermeir, I. and Verbeke, W. (2006) 'Sustainable food consumption: Exploring the consumer "attitude-behavioral intention" gap', *Journal of Agricultural and Environmental Ethics*, vol 19, pp169–194
Warde, A. and Martens, L. (2000) *Eating Out: Social Differentiation, Consumption and Pleasure*, Cambridge University Press, Cambridge
Wilk, R. (2010) 'Consumption embedded in culture and language: Implications for finding sustainability', *Sustainability: Science, Practice & Policy*, vol 6, no 2, pp38–48
Willer, H. and Kilcher, L. (eds) (2010) *The World of Organic Agriculture, Statistics and Emerging Trends, 2010*, IFOAM and FiBL, Bonn and Frick
Willer, H. and Yussefi, M. (eds) (2004) *The World of Organic Agriculture: Statistics and Emerging Trends, 2004*, IFOAM, Bonn
Willer, H., Yussefi-Menzler, M. and Sorensen, N. (eds) (2008) *The World of Organic Agriculture: Statistics and Emerging Trends, 2008*, IFOAM, Bonn
York, R., Rosa, E. and Dietz, T. (2003) 'Footprints on the earth: The environmental consequences of modernity', *American Sociological Review*, vol 68, no 2, pp279–300

Young, W., Hwang, K., McDonald, S. and Oates, C. (2009) 'Sustainable Consumption: Green consumer behaviour when purchasing products', *Sustainable Development*, vol 18, no 1, pp20–31

# 12

# Conclusion

## Introduction

This book has examined the globalization and sustainability of food provision. With the help of several illustrative cases, it has become clear that securing sustainable food provision under present conditions presents a significant challenge. To help readers better understand the full nature of this challenge, as well as possible solutions, we introduced several conceptual tools that have been developed within the social sciences. Together, these tools help shed light on entrenched practices and views in food provisioning. Throughout this volume, when analysing these developments, global and local dynamics have been combined, as both are essential for understanding and moving forward. Findings from the natural sciences have been included as well, to provide additional depth in addressing the difficult problems that we face today in developing the sustainability of our food supply.

This concluding chapter summarizes the most important results from this volume, discusses the future of food provisioning, and makes a closing argument on the value of social-science perspectives for achieving greater sustainability in producing, trading, processing and consuming food in the context of global modernity.

## Primary Findings

This volume has presented an overview of the main trends in contemporary food provision related to globalization and sustainability, introduced key social-science concepts for analysing these trends, and connected these trends and concepts with ongoing societal and political debates about food provision and its future.

## Globalization

Chapter 2 sketched the main characteristics of the process of globalization in food provision. Since the 1980s, food production and consumption have become increasingly distanced in both spatial and social respects, bringing with them all kinds of unexpected challenges. Addressing these challenges, including those related to the sustainability of various food-related practices, first requires an in-depth understanding of the processes themselves. Among the different conceptual frameworks that are available for this purpose, in our view, the sociology of networks and flows offers promising perspectives. Using the concept of networks, we can link the multiple social actors who are directly and indirectly involved in food provision. These actors are engaged in numerous food-related social practices organized at various locations and on different levels. The notion of flows helps explain how food is connected to these different social practices, because food flows entail not only the material aspects of food from 'farm to fork', but also the related financial, cultural, political and informational/communicative dimensions that can be associated with these material flows. With the help of these concepts, it becomes possible to explain how particular local food-consumption practices influence food-production practices at large distances (and vice-versa). We can also apply this conceptual framework to explain that even if the food itself does not travel over long distances, its production methods, quality requirements, etc., are still highly influenced by global processes, and how, at the same time, particular local characteristics of production and consumption practices influence global dynamics. When introducing this conceptual framework, we underlined the importance of recognizing that the organic character of food prevents it from being completely disconnected from local conditions. Food remains, at least for the time being, bound to particular local dynamics. In matters of food, global and local dynamics are therefore indissolubly connected, although not necessarily congruent. As a result, a fundamental tension exists between the way food is integrated in the emerging global and virtual 'spaces of flows' and the remaining local and organic specificities in the 'spaces of places' where it is grown and consumed.

This conceptual framework of food provision in global modernity as comprising networks and flows enables us to combine social and material aspects, as well as local and global dynamics, when analysing problems of sustainability.

## Sustainability

Sustainability of food provision, the subject of Chapter 3, remains difficult to define. At the broadest level, sustainability – understood as assuring the optimization of economic, social and ecological interests of present and future generations – is widely accepted as the overarching aim for organizing food provision. Further detailing and elaborating this general definition has proven to be problematic, however. Many different definitions have been suggested, some very broad and others very restrictive, but no unanimous understanding of what sustainable food provision entails has yet been reached. Over time, definitions have changed and new elements have been included, so the exact definition of 'sustainable food' will probably remain an issue of debate in the future.

Nevertheless, some general trends in how sustainability in food provision is understood can be identified. Over time, environmental and developmental concerns have become more integrated, while sustainability is also becoming more than a matter of agricultural production alone, but is nowadays seen as an issue for the food-supply system in its entirety. Promoting sustainability involves the production, processing, trade and consumption stages of food provisioning, and therefore requires an integrated approach. These trends also relate to the process of globalization because overall awareness has grown about sustainable food provision being a global challenge, making it everyone's responsibility. At the same time, globalization also transforms the spreading of environmental and food-safety risks because distant locations may now be connected through the transport of agricultural or food products, while individual governments are unable to fully control these movements on their own. At the same time, innovations in global communication and transport technologies and increased opportunities for global travel are improving opportunities for more closely connecting food producers and consumers, although their physical distance may increase.

As a result, different pathways towards increasing sustainability emerge. Improvements can be achieved in how food is grown on farms, as well as during the processing, transporting and consumption phases. Each phase encompasses different institutions, engaging multiple actors in various ways. Governments, markets and civil society all present different modes for adapting food-production practices, internalizing environmental impacts, and engaging producers and consumers in moving towards sustainable development. To be effective, these institutions and social actors should ensure that appropriate feedback mechanisms are operational and organized on the appropriate scale. For instance, in larger-scale food supply systems, such mechanisms should be able to handle geographical and temporal distances and operate at a time and rate that is suitable to elicit the correct adjustment. This requirement explains the rapidly growing number of innovative governance arrangements, such as food labels and private certification schemes, which try to bridge the gaps between producers and retailers and between producers and consumers. Other pathways to increased sustainability in food provision focus primarily on reducing the physical distance between producers and consumers by connecting them directly wherever possible and limiting the food supply's processing phase.

All of these and other pathways to promote sustainable food provision converge on the need to take an integrated approach in rearranging the supply chain from producer to consumer and from farm to fork.

## Regulating food

Globalization is creating unfamiliar challenges for regulating food, especially when the aim of increasing sustainability is taken into consideration. In Chapter 4, we elaborated on this issue by discussing the limits of conventional, government-based modes of regulation under the present conditions and by reviewing different alternative governance arrangements that are being introduced.

Conventional regulation of food was based on policies and legal measures introduced by sovereign nation states to protect their populations and to secure their economies and environments. This approach is becoming increasingly inadequate under the conditions of accelerated globalization because national governments are unable to effectively control food that is internationally traded or address food-related global environmental problems on their own. International food trade has grown to such proportions and to such a degree of complexity that traditional public institutions lack the capacity and sometimes also the authority to assure the safety and sustainability of food.

In reaction, many governments seek collaboration with other governments. Through multilateral agreements and transnational authorities, they intend to address cross-border issues, such as regulating international trade through the WTO or global climate change through the UNFCCC. Such transnational networks may address common problems, but they are often hampered by slow decision-making, as unanimity is required, and by their basic principle of respecting the sovereignty of each government. Complex, sensitive and controversial issues such as sustainability in food provision are particularly difficult to incorporate into transnational decision-making mechanisms and to translate into effective multilateral agreements.

This situation has made other societal actors, such as NGOs and private companies, engage in governing food under the conditions of globalization. This process is framed as the transition from 'government' to 'governance'. Hence, at present, food governance simultaneously involves national governments, transnational institutions and multilateral agreements, local authorities and a wide range of private initiatives. Each of these different governance actors has their own objectives, responsibilities, strategies and ways of operating, while being active at different levels. As a consequence, we now are faced with a multitude of coexisting, and sometimes even competing, food-governance arrangements. Although national governments continue to play a critical role in many of these newly introduced food-governance arrangements, other social actors, including private companies, consumers and NGOs, take up more responsibilities.

As a consequence, the different actors in the food-supply chain are faced with numerous official and private schemes, regulations and labels, many of them dealing with issues of sustainability. Harmonizing the different public and private food-governance arrangements is a challenge, as there is no recognized authority to effectuate this, nor is there unanimity on the final aims of the resulting global sustainable food-governance arrangement. Multiple private labelling and certification schemes thus are likely to continue in the future, particularly as private food-governance arrangements seem more flexible and have the opportunity to address particular societal concerns that are not (yet) part of national regulations.

## Climate change

Chapter 5, the first of four chapters discussing particular cases, dealt with the relationships between global warming and food provisioning. Applying the concepts and referring to the debates presented in the preceding three chapters, climate change can be considered as a prime illustration of how closely globalization, sustainability and food are interconnected. Global dynamics, such as those related to population growth, rising incomes, increased use of energy-intensive technologies in food production and processing, long-distance air transport and changing diets, all affect how food provision contributes to climate change. At the same time, agriculture also suffers the consequences of global warming, as conditions in fragile areas, such as sub-Saharan Africa, may become too harsh or too volatile for producing food. Therefore, there is a need to take food provision into account when addressing climate change and to adapt present-day agri-food-supply practices to reverse this growing environmental risk and accommodate its possible consequences.

In recent decades, various consumer-oriented strategies have been initiated to reduce the climate impact of food provision. Labelling products on the basis of 'food miles' and 'carbon footprints', which is intended to influence consumers' buying behaviour, is a clear example of this. Food miles were an early tool aimed at reducing the overall distance a food product travels from producer to consumer and at promoting the consumption of locally grown food. Over time, however, this relatively simple tool became controversial because critics argued that its effective application would threaten the export opportunities of farmers in less-developed countries. Other exporters tried to show that the $CO_2$ emissions from food transported by ship to Europe were rather insignificant. Therefore, carbon footprints were promoted as a more reliable and refined alternative for guiding consumers' choices. Carbon footprints are based on the actual measurement of the climate impact resulting from a particular food product during its life cycle. Carbon footprints also proved controversial, however, because in practice, standard data must be used that are too general to generate the dynamics needed to move towards more climate-friendly food production practices. Furthermore, these labels may lead to radical shifts in the geographical distribution of where food is produced and processed, including reducing export opportunities for farmers in Africa. Many consider this to be an unintended consequence because it places the burden of combating climate change on people who were not responsible for creating it.

The information provided through carbon labels may nevertheless serve as an appeal to supply-chain actors, such as agri-food companies, retailers, consumers and governmental authorities, to change their behaviours. Such labels can be considered a useful informational tool, but unless carbon labels are embedded in wider transition processes, it is unlikely that people will change their behaviour. Also, it is often unclear who owns the label and how decisions are made on how to measure the indicators. In combination with the presence of various schemes, this lack of transparency is especially problematic. Without an international consensus between the different national authorities and relevant private actors, it is unlikely that carbon labels can be applied as an effective governance tool for promoting sustainability in the context of globalizing food provision.

## Localizing food provision

Localizing food supply in response to the problematic aspects of globalized and industrialized food provision, the subject of Chapter 6, has become popular among many consumers and social movements. Through the creation of short supply chains, various social actors aim to strengthen food sustainability because more fresh and seasonal foods can be produced and consumed locally. Such alternative food-supply chains are expected to demand minimal energy for transport, processing and packaging and to maximize the freshness and quality of the final food product. When producing local foods, agricultural production techniques can be applied more in concert with local ecologies and tastes, and in doing so, they arguably reduce environmental impacts. Local supply chains furthermore optimize available diversity, reduce energy needs and prevent problems of oversupply or scarcity by promoting direct communication between producers and consumers and restricting the role of market dynamics in the food supply.

Broad support is given to various attempts to re-establish closer relationships between food producers and consumers. Social movements promoting local agri-food supply chains bring together many people concerned about the consequences of the contemporary way of industrialized food provision. At the same time, the local food movement constitutes a heterogeneous coalition of different groups and concerns, including farmers and activists fighting corporate farming and protecting local communities, consumers looking for fresh and healthy foods of high quality with low environmental impact, and concerned citizens trying to reduce malnutrition (including obesity) among poor people without adequate access to healthy food. In this coalition, activists aim to address such issues as food quality, food-related health issues, environmental impacts, social justice and the ethical dimensions of contemporary globalized, industrialized foods. These multiple concerns cannot all be expressed in a single model of local food provision. Local food-supply chains therefore can acquire various organizational shapes and forms, and there are few formal standards and procedures.

It is important to stress that local food provision cannot be understood as being completely separate from global food provision. Many different relationships and influences exist between them, whereby knowledge, technologies, impacts, etc. are all involved. Even beyond this level of direct influence, local identities are not fixed but flexible and mouldable because such identities must be created and recreated actively by the social actors involved. The local identity is a site of resistance against the global, but the site itself is created through the process of globalization. Therefore, it may be unhelpful to romanticize the local aspect of food supply systems and ignore the continued relations with wider levels and actions.

## Fair trade and consumer trust

Fair trade labelling of globally traded food products was selected as the third concrete case, and was the subject of Chapter 7, because it may contribute to increased sustainability in globalized food provision. Fair trade is an effort to build more equitable relationships between producers and consumers. Since its introduction, the fair trade movement has become a frontrunner in introducing market-based tools to mobilize consumers. Through the introduction of producer and retailer standards and product labels, consumers are equipped with concrete tools to operate in markets not only on the basis of economic principles but also through applying non-economic values, such as equity and sustainability. Fair trade bridges the gap between the demand for more social justice and environmental sustainability in the global food supply and the limited facilities provided within conventionally operating markets for implementing these aims. A crucial intention of fair trade is strengthening the position of Southern producers in global food-supply chains through concrete support from a civil-society movement. As a result, an innovative constellation has emerged where civil actors, private firms, producers and consumers incorporate other, non-economic values in economic relationships.

The fair trade movement tries to establish and maintain consumer trust in its label and its strategy, but this is a permanent challenge because this abstract, institutionalized trust cannot be taken for granted. First, abstract trust is permanently vulnerable to criticisms and negative examples. Second, with its success, fair trade must be marketed not only to a restricted group of committed consumers, but also to more mainstream consumers, who may insist on getting bonuses from buying fair trade-labelled food products in addition to supporting small-scale farmers in developing countries.

In the last decade, the fair trade movement has been extremely successful in promoting its products, as its sales in many OECD countries have experienced high growth rates. This success has inspired other initiatives that may entail different (consumer) concerns but nevertheless apply the same strategy of using consumers' buying power to realize changes in global food-supply chains. Still, although the fair trade movement increasingly seeks to address the environmental impacts of global food trade as well, the initiative remains unique in its aim to change the socio-economic position of food producers in developing countries. This socio-economic aim also makes the fair trade movement vulnerable to critique because it sometimes may be difficult to attribute success (or failure for that matter) to only its certification and labelling practices when farmers are incorporated into global markets and part of local communities at the same time. Therefore, it may be necessary to supplement the fair trade labelling of food products with more encompassing strategies that also take local production circumstances and global commodity markets into account.

## Sustainable fish provision

In Chapter 8 we discussed the case of the global fish supply. More than most other food products, fish has become a definite part of food networks where both global supply and (particularly ecological) sustainability concerns play important roles. The growing demand for fish poses serious threats to the survival of many species, and therefore, effective measures are urgently needed to manage fish stocks. National governments have tried for many years to reduce the environmental pressure from fisheries, but they have had serious difficulties in implementing effective measures because of the fishery industries' vested interests and the challenges associated with creating effective policies across the different seas and oceans. Governments have introduced stringent fisheries quotas and marine protected areas and tried to combat illegal, unregulated and unreported fishing. Nevertheless, over the years, the number of fish stocks under pressure has only increased. In reaction, environmental social movements have supplemented their more traditional lobbying and awareness-raising campaigns with different innovative arrangements that try to secure the future sustainability of fisheries' resources more directly.

These innovative arrangements include certification and labelling schemes, as well as fish wallet-cards that try to make reliable information about the (lack of) sustainability in particular fish provision activities available to the consumer. Together with consumers, these environmental NGOs consider themselves to be more effective co-regulators of fisheries management and drivers for change than nation state agencies or private companies. These environmental NGOs can operate at the global level and are better able than national governments to cross national borders. They may even trespass on the sovereignty of individual nation states when they try to implement parts of their global sustainability agendas by using consumers' buying power. Using this flexibility, and calling upon science as their main source of legitimation, these environmental NGOs seem capable of creating pressure on unsustainable practices.

The experiences of private actors in preventing and reducing the environmental deterioration of this important global natural resource provide interesting learning experiences for other organizations and products. Environmental social movement-initiated certification and labelling schemes and different consumer-oriented strategies may not necessarily replace governmental arrangements, but they offer important additional instruments to secure the availability of fish as a source of healthy food for future generations. Their main contributions are their flexibility; the opportunities they offer for actively involving different actors, especially consumers, as co-regulators in managing global fish networks; and their particular advantages of creating trust among consumers.

## Future roles of producers

Chapter 9 was the first of the final three chapters, which discussed the future of global food provision from the perspective of different social actors. This chapter dealt with the roles producers can play in creating this future, with particular attention to the position of smallholder farmers. Food producers may engage in multiple strategies to strengthen their positions in different configurations, combining market opportunities, technological innovations and consumer and sustainability concerns. There is no single strategy for farmers when trying to contribute to an agriculture that reduces hunger, increases sustainability and strengthens the position of its producers under conditions of globalization. Multiple fits between these aims and the specific conditions of producers are possible, resulting in different strategies. Trying to make a general distinction between strategies that remain within the conventional mode of mainstream farming and those that can be considered alternative is not very helpful because all food production practices are embedded in wider social networks and not permanently fixed. So-called conventional farming practices also evolve and may present more sustainable alternatives at higher levels, while hybrids or mergers of different strategies can be imagined as well. In a similar vein, technological development should not be excluded either. Technological innovations may deteriorate the environmental performance of agri-food supply chains, but they may also promote its sustainability, provided this aim is already included in the development stage and potential side effects are taken into account.

## Restructuring the food supply: Supermarkets and sustainability

Chapter 10 illustrated and analysed retailers' key role in global food-supply networks. Since the 1970s, in many countries, the central position that producers and traders traditionally held in food-supply chains is being taken over by large retailing corporations. Today, different large retailing firms operate beyond the control of national authorities. Retailing firms increasingly are imposing health and safety conditions beyond official legal requirements and specifying food-quality standards; they are having a huge influence on the terms of trade under which producers must operate.

At the same time, there is considerable variation among the different retailing companies and their strategies, including how they deal with matters of sustainability. An important cause of this variation is the dilemma that retailers face in terms of their involvement with the food supply itself. By increasing their control over suppliers, engaging directly with producers and processors, retailers may have easy access to the preferred food products with the required characteristics and at the scheduled moment. This strategy, however, greatly reduces the flexibility that retailing companies prefer when operating under the present dynamic market conditions, where outsourcing seems much more preferable. A compromise between these competing demands is the development and introduction of private standards and certification schemes, such as

HACCP, that can be applied throughout the sector by all suppliers. Such standards create a level playing field for supermarkets guaranteeing precise product qualities and the use of production methods, but still allow for flexibility in their operations.

Some retailing corporations may find it attractive to include sustainability goals and principles in their strategies because they create interesting economic opportunities, strengthen their public image and reduce their operational expenses. Selling more sustainable food products may attract consumers who are willing to pay more for such foods, allowing retailers to profit from the higher margins on quality foods. Being involved in promoting and selling more sustainable products may protect retailers from becoming victims of public 'naming and shaming'. Retailing firms may also express genuine commitment to the principles and goals of sustainable development.

Whatever their motives, retailer involvement in transitions to more sustainable food provision is essential because of their key position in global food networks. Such involvement may be more easily realized when direct consumer-related issues are concerned, such as food safety, sustainability and animal welfare, than when the stated goal is improving social conditions for producers. Providers may initiate steps towards green-food provision themselves, but governments or consumers, in collaboration with social movements, may also motivate such steps. Essential here is how sustainability concerns are interpreted, put into practice and communicated with producers and consumers. So far, there is little real understanding of how the scientific, technological and economic rationality of the providing system can be translated in the more subjective, social and value-driven rationality of the consumer, and vice-versa. A better understanding of these mechanisms may greatly improve the effectiveness of sustainable strategies in the context of global food provision.

## Consumer involvement in sustainable food provision

Consumers, the subject of Chapter 11, are essential in global food networks, and should therefore not be ignored. In different ways, consumers may be constitutive for changes in global food provision. They react to developments within supply chains, such as the introduction of food products with additional sustainability claims or changes in the organization of the food supply itself. Consumers may also express their sustainability concerns through changing their buying behaviours or through pressure via social movements or the government.

Understanding these dynamics requires a social-practices approach to consumer–supplier interactions. Although not every consumer behaves the same, not every consumer acts completely differently either. A social-practices approach allows us to find a middle way between these opposites because it considers consumer behaviour as being embedded in wider social structures.

The role of food consumers in bringing about sustainable development should therefore be studied in the concrete everyday practices through which they buy, prepare, eat and discuss food. Each of these social practices has a particular constellation of social actors, discourses, (routinized) behaviours, balances of power and presence of (different forms of) trust. Today these practices acquire their specific shape in the context of globalized food provision, which transforms the conventional roles of social actors and brings in unfamiliar challenges about safety and the environmental and social impacts of food. As a result, different consumption practices emerge, whereby some consumers buy their food products with sustainability labels in their familiar supermarket, others buy more sustainable food products via devoted specialized distribution channels (organic food shops, farmers' markets, direct supply systems) or change their diets (becoming (part-time) vegetarians). At the same time, many consumers still hold on to their conventional routines in buying and consuming food, expressing different arguments for not being willing to change their habits.

When trying to understand these consumer dynamics, the eventual changes in buying food should be related to wider social processes through which actors such as NGOs, community groups, social movement organizations, politicians and the media are involved in discussing and advocating sustainable food provision in global modernity.

## An Agenda for Food Sustainability

As this summary shows, promoting sustainability in food provision under conditions of global modernity is both a conceptual and political challenge. It is no longer possible to take food provision for granted, as has been the case for many years in more affluent societies. We can no longer be assured of agriculture's continued capacity to supply sufficient food for a growing global population.[1] Agriculture and food now figure much more prominently on the global political agenda than was the case for several decades.

Many governments, specialized institutions and policy advisers agree that there is an urgent need to reconsider conventional models of food provision. They acknowledge the fundamental changes in the globalizing of food markets where new food superpowers (Brazil, China and India) are emerging and where the private sector is consolidating into a limited number of very large transnational agribusiness companies (Foresight, 2011). At the same time, global production is becoming more volatile as a consequence of adverse weather conditions, the use of limited available resources for the production of bioenergy, speculation on the commodity markets through financial instruments such as *futures* (see Ghosh, 2010), and a growing demand for food in China and other rapidly growing economies,[2] often in combination with changing diets.[3,4] The context for interventions is even further complicated by the growing seriousness of several environmental problems (Lang, 2010), such as the dwindling availability of freshwater resources, the threatened (agro-)biodiversity and ecosystem services and the decreasing abundance of energy from non-renewable sources.

Through these developments, governments, specialists and social movements are faced with multiple issues that they must address simultaneously. They must address these issues in a context with many uncertainties regarding the adequacy of the present international institutional architecture to respond to future threats to the global food system; in particular whether there is sufficient political will to enable these multilateral institutions to function effectively is yet unclear. In reaction, various public and private actors feel forced to reconsider their policies and formulate agendas for the future. Among the many food agendas or policy programmes that have been presented recently, the proposal for a new CAP for the EU and the different agendas for addressing food and climate change stand out. We therefore present and discuss these agendas below in some detail.

## The future of agricultural policy in the EU

The EU has supported its farmers for decades. Although this support has remained quantitatively the same for a number of years, it is shifting from an orientation on supporting farmers by controlling agricultural production and guaranteeing their price towards trying to support farmers' incomes without distorting trade. Therefore, instead of organizing interventions intended to secure the commodity price, farmers are nowadays supported directly through payments that are decoupled from their actual production.

In 2010, the EU initiated a review process to allow for the implementation of an improved agricultural policy in 2014. In a proposal, the European Commission (EC, 2010) formulated three main objectives: viable food production, sustainable management of natural resources and climate action and balanced territorial development (see *Bridges Weekly*, 2010). The EC argues that the EU needs to maintain its capacity to produce food, but at the same time respect its commitments to promote international trade and maintain policy coherence for development.[5] In concrete terms, this policy orientation means that overall, the market orientation in the CAP is strengthened and that farmers are offered less income security than in the past (Matthews, 2010).

The resulting public debate on these proposals addresses two main sustainability issues: the ecological and social consequences of the new CAP.[6] For instance, in its alternative proposal for the future CAP, the French Ministry for the Environment (Ministère de l'Ecologie, 2010) defined three core principles:

- a guarantee of food and environmental security and coherence between citizens' demands and the products offered to consumers;
- a greater equity in income support and a better balance with payments for ecological services;
- a general reorientation of the systems of production towards sustainable farming (agriculture agro-écologique) and stronger protection of the environment.

In this proposal, the economic dimension when producing food is greatly reduced and replaced by an orientation on environmental security, with the understanding that recognizing ecological limits when producing food will better protect the environment, secure the future of producing food, and increase income opportunities for the rural population by increasing payments for ecological services.

This approach is in line with an alternative policy agenda presented by the Agricultural and Rural Convention (ARC, 2010).[7] ARC (2010, p2) suggests as aims for the future CAP:

> *food security, a fair return to farmers, food quality and public health, sustainable standards in agriculture, land security, holistic protection of the environment, mitigation of climate change, strengthening and diversification of the rural economy and the well-being of rural communities.*

According to this NGO, securing food for the growing global population should not be achieved by concentrating food production in some limited, highly productive, regions and relying on massive food transports to distribute the harvest. Rather, they argue for promoting a high degree of self-sufficiency and food sovereignty at local, regional, national or continental levels. The EU should therefore also produce most of the basic food commodities it needs and broadly limit imports of food or feedstuffs. In its CAP, the EU should also secure its farmland for long-term use by promoting sustainable management.

In the debate on the future of the CAP, the EU's focus seems to be primarily on how to combine the economic objectives of maintaining a viable agricultural industry while controlling the budgetary consequences, with the social and environmental goals of securing food for the European population and reducing the negative environmental consequences. Global dynamics seem to hardly play a role in these discussions; they are only a limiting condition and not an essential building block.

## Food-policy agendas on climate change

Growing concerns around the world about the impact of climate change on the sustainability of agriculture are encouraging many research institutes, NGOs and experts to reflect on the direction of future food policies.

The IFPRI, for instance, argues that coordinated interventions are urgently needed to prevent climate change from becoming a threat to the food security of the poor (Nelson et al, 2010). Next to the familiar objectives of increasing poor people's incomes and reducing GHG emissions, the IFPRI proposes to invest in agricultural productivity improvements to mitigate the impacts of climate change, enhance sustainable food security and strengthen international trade arrangements to compensate for different climate-change effects in various locations.

More focused interventions to improve agricultural productivity in Africa are formulated by Hunt and Lipton (2011) in a report for Chatham House (UK). They suggest focusing on restoring and protecting soil fertility and on reducing water use by developing appropriate technologies and increasing low-external input and organic farming. Furthermore, they propose that smallholder farmers be supported in securing their land tenure, improving their access to appropriate credits, extension and farming inputs, facilitating their access to markets by improving infrastructures, and making markets more transparent and liberal.

These proposals illustrate the need for engaging with the consequences of climate change for food production by addressing both technological and political dynamics. Nevertheless, while they stress the urgency of taking action, these agendas remain rather conventional in their content and actual recommendations. If the conclusions from the various chapters in this volume are taken into account, there may be a need to extend this debate and consider alternative policy frameworks and pathways of technological change.

## Policy in the future food agenda

The agendas presented above are quite illustrative of the present state of the debate on future food-provision policies. They largely rely on national (or international) governments taking policy measures, eventually coordinated through multilateral agreements. The recent global financial, climate and fossil-fuel resource crises show that conventional domestic policies are no longer adequate in addressing such cross-border problems. The future of food provision is another one of these problems, and it faces similar limitations.

Although solutions are not simple, it is evident that governments, politicians and civil-society organizations must rethink their roles and strategies under conditions of global modernity. Governments in poorer countries often lack the means to respond to emerging challenges, and while governments in richer countries respond, they mostly focus on finding short-term solutions only and fail to address more fundamental issues. Richer countries must consider their way of interacting with newly emerging powers, not only by focusing on economic dynamics, but also in terms of different global cultural, social and political trends. Governments must become aware that they can no longer operate autonomously but need to function in a network of states in global modernity.

Our global society is increasingly influenced by transnational communities of non-governmental actors and businesses organized at multiple levels, including locally (Bieckmann, 2011). Among them, civil-society actors cannot be ignored; their political force lies not so much in representing the mass public, but rather in their ability to convince the media and retailers that they speak to and for the critical public. This latter, leading segment of the public is essential to food retailers' sourcing practices and sales (Freidberg, 2004). To succeed in transnational food-provisioning campaigns, NGOs must connect global relationships and local dynamics: the (critical) consumers must be convinced to support their claims.

Functioning food markets are indispensable for global and local food provision. As the majority of the world's population now lives in cities, individuals rely primarily on markets to assure their access to food. These markets often do not operate transparently, however, and are often unfair and unfavourable to the poor.[8] Under the conditions of globalization, market power has shifted from large traders to retailers and suppliers of input (Murphy, 2006), but the weak position of small producers and individual consumers has remained. Governments could try to correct this power imbalance, but by doing so, they may focus more on consumers' than on producers' interests. Most governments nowadays, however, are primarily concerned with controlling their budgets and about the health outcomes of dietary choices for their population in general. To correct the power imbalance also in the interest of producers, global coordination between different governments is needed, but it is not evident how this should be arranged and where the appropriate policy instruments can be found. As national agricultural policies did not completely control the behaviours of individual farmers, global food policies can also not be expected to control concrete activities, but they could still provide a framework for the orientation towards activities that different social actors should take. National food policies need to be replaced by a shared global policy, involving networks of public and private actors, towards securing food as a global public good.

## Technology in the future food agenda

For more than a century, national agricultural policies have focused on producing more food. Especially since the 1960s, governments and specialists expected much from technological innovations to increase agricultural productivity, for which the Green Revolution (see Box 3.4) has become an icon. Since the 1980s, however, the interest in agricultural technological innovations has diminished, and this trend needs to be reversed, as it requires more attention.[9] The conventional model of simply adding more science and technology needs to be replaced by more encompassing paths of technological development.

Today, the application of available technologies could still increase average yields substantially, particularly in Africa and the Russian Federation. Nevertheless, still further technological advancements are needed to supply the growing global population, and therefore, its orientation needs to shift. After the successful 20th-century Green Revolution, there is need for a 21st-century version. Or, as Olivier De Schutter, the special rapporteur on the right to food, declared in his report to the UN, 'pouring money into agriculture is not enough, what is most important is to take steps that facilitate the transition towards a low-carbon, resource preserving type of agriculture that benefits the poorest farmers' (De Schutter, 2010, p3). A revised programme for technological development therefore should focus not only on increasing yields, but also on shifting the model of agriculture from one that is *input*-intensive (in water, fertilizer, pesticide and energy) to one that is *knowledge*-intensive. Thus, there is not a singular pathway for technological development because different systems of

## Box 12.1 Integrated agricultural development for Africa

When identifying possible ways forward for farming in Africa, Andriesse et al (2007) concluded that focusing only on technological developments to increase agricultural production would not address the structural causes of food insecurity on the continent. It is important to get the basics in place and facilitate farmers' access to the tools they need to increase their production and to profit from the results. Increased agricultural productivity needs to be integrated with providing access to resources and services and with markets that work for the poor. Farmers do not operate in a vacuum, so to make this process possible, these three domains cross-cut the domain of improving institutional arrangements that affects all other domains (see Figure 12.1).

| ***Increasing productivity*** | ***Providing access to resources and services*** | ***Making markets work for the poor*** |
|---|---|---|
| *Institutional arrangements*<br>Formal and informal<br>at local, national/regional and international scales | | |
| Developing and applying pro-poor and sustainable technologies and production systems that enhance land and labour productivity and economic returns | Providing producers/ local entrepreneurs/ rural communities with equitable and secure access to natural resources and livelihood sevices | Creating trade and market conditions that enable small scale/ poor producers and rural entrepreneurs to access and participate in markets |

**Figure 12.1** *Model for integrated agricultural development*

*Source:* Andriesse et al (2007, p54)

More concretely, Evans (2009) identifies the key resources that farmers need: assets (such as land, machinery and renewable resources), markets (adequate infrastructure, communication networks and information), credits (to allow smallholder farmers access to necessary inputs), knowledge (extension services, and research and development) and risk-management tools (social protection, insurance and crop storage).

Any technological innovation must fit into these complex dynamics, especially when they intend to contribute to sustainable development.

food supply exist and each of them requires technological innovations that fit its particular conditions.[10] Still, most developing countries' agricultural research is oriented towards increasing crop productivity, disregarding the potential of smallholders and their particular conditions. Focusing more on integrated research based on an acknowledgement of the complex farming systems of smallholders would probably yield more effective technological innovations. Such research would require better-targeted publicly funded research and more active stakeholder participation (see Box 12.1).

Technological change in food provision has concentrated on increasing agricultural productivity and securing sustainability, but innovations also occur during the food-processing stage. These developments, in turn, have an impact on primary food production. One such consequence is the strengthening of the bifurcation process in food production. On the one hand, high-valued quality food items, such as fresh fruits and vegetables and fish, are produced and directly sold through the market, while on the other hand, agricultural commodities are produced as inputs for industrial processes that generate a wide variety of end-products, including food.[11] These different end-uses directly influence the kinds of technological innovations that are targeted. While quality, safety and sustainability are prime characteristics in organizing the first category, technological change in the second category is essentially geared towards more efficiency, uniformity and low processing costs.

Decisions on technological developments imply 'social and political choices, in a contested space, within which different interest groups advance particular arguments' (Foresight, 2011, p16).[12] Each innovation is also associated with certain positive and negative externalities and cannot be decided only by technological experts. So, while Borlaug may argue that we have the technology 'available or well advanced in the pipeline' (2000, p490) to feed a population of 10 billion, but 'extreme environmental elitists' are derailing its application, this denigrating comment does not make environmental concerns disappear. Therefore, there is need for technological improvements to increase the production of food and to promote efficiency in the use of scarce resources, but these developments should not be isolated from wider social and institutional dynamics (Vellema and Danse, 2007).[13] Next to the directly involved groups, consumers, local institutions and global governance arrangements also should be engaged when discussing technological innovations in agri-food.

## The Contribution From the Social Sciences to Future Food Agendas

The future food agendas presented above focus on their political and technological dimensions, but this book argues that wider contributions from the social sciences should be included as well. In the context of global modernity, food policies entail many social actors whose involvement in different social practices is little understood, while technological development also means social change.

Global modernity raises multiple challenges that need to be addressed. Dealing with these complex questions needs strong public institutions, but these are often not available, as many governments are only getting weaker. In reaction, authors such as Lang (Lang et al, 2009) and Timmer (2008) insist on the need to develop a new paradigm for the future of food: a holistic approach is necessary because all problems hang together in some way. They propose to replace the productionist paradigm of the past with a new one that aims to 'feed everyone sustainably, equitably and healthily'. Future food systems have to be 'diverse, ecologically sound and resilient in the face of increasing environmental, economic, or social volatility and create robust and sufficient supply systems and stocks' (Lang, 2010, p95). The principles and mode of operation of these systems should be such that they can be 'maintained for the long term, thereby enhancing – not just protecting – the land's productive capacity; and build the capacities and skills necessary for future generations' (Lang, 2010, p95). This description of an alternative future food system is attractive, but its elaboration and implementation requires a better understanding of what these different concepts actually mean, which actors are involved in what ways and how different opinions and interests can be addressed. Responding to these questions requires contributions from the social sciences that are capable and prepared to study everyday social practices of producing, processing, retailing and consuming food and the different institutions involved in structuring these practices.

For instance, analysing alternative global agri-food networks, such as fair trade and global organics, needs a re-conceptualization of the notion of quality and an understanding of the changing relationships between producers and consumers. A general judgement on their success or failure without detailed empirical research may not be very helpful. Quality of food then proves to be more than a set of simple product characteristics but is expressed through social networks that are fair, sustainable and democratic. The collaboration between producers and consumers as social partners, as well as economic agents, allows for combining better prices and goods with meeting social and environmental standards. Only empirical studies will enable us to understand whether and how declared aims are being realized in individuals' everyday lives.

As mentioned, unanimity about the future of food is unlikely, as there are conflicts of interest, differences in power, multiple roles in food provision, etc. From the perspective of global food provision, it is essential to include developing countries, small-scale farmers and poor households in these debates and to recognize that defining sustainability will remain a permanent challenge. Social science can make essential contributions when analysing and trying to understand these debates and conflicts and perhaps also in addressing them.

## Final Remarks

Intensified economic globalization involves complex challenges for feeding the world's growing human population in a sustainable manner. More food must become available to more people, and it should be produced in more sustainable ways than it is today, but these goals will have to be realized under unfamiliar conditions. Many conventional strategies, based on national government policies, markets and technological innovations, have lost much of their effectiveness. This book has illustrated these challenges and introduced a variety of innovative concepts and practices that can contribute to analysing, developing and implementing the future provision of food.

The future provision of food will probably not be integrated into one single coherent system but, rather, will result in a patchwork of different practices. Some of the newly introduced practices may be highly integrated in global supply chains, and others will be more self-organized; sustainability of both globally and locally based food systems can be improved with the help of various technological approaches. The possibility of multiple, hybrid pathways to more sustainable food provision needs recognition, without ignoring that it remains important to analyse who is dominating their agendas.

Ultimately, we must acknowledge the agency of all people involved in food provision. Social-science perspectives can contribute to increased sustainability by showing that beyond laboratory-tested technological innovations and abstract economic models, real people construct the future of food in their everyday lives. Their efforts need recognition and support to make the food supply more sustainable from a global perspective. A sustainable food future requires a democratic global management of ecological resources. In the end, food will remain a unique commodity because of its organic character and its socio-cultural embeddedness in everyday life; to be successful, policy approaches for food sustainability must take these special qualities into account.

## Notes

1 For many years, the growth of food production was expected to continue without much additional effort. For instance, agriculture constituted 17 per cent of international development aid in 1980, but shrank to 4 per cent by 2000 (*New York Times*, 2009).

2 The world's population is still growing and is expected to rise another 25 per cent by 2050, while urbanization also continues to be rapid, thereby putting pressure on rural populations to produce enough food (Lang, 2010). Securing the labour available for farming in the future may also be a challenge, as the position of people working in agriculture is weak in terms of income, social security and labour safety. This may lead to difficulties in attracting qualified workers for more technology-intensive farming.

3 Diets may change and lead to public-health problems when they shift from simple staples to more high-value-added processed food products, increasing human health risks, such as obesity.

4 Another contributing factor to volatility on the global market is foreign interests' increasing control of land areas for food production through land purchase and leasing agreements (Foresight, 2011).

5 It is not clear what this proposal will mean for farmers exporting from developing countries and whether the changing (but nevertheless continuing) support to European farmers will lead to more favourable conditions for them. Matthews

(2010) refers to the shift in the EU's trade status as it has changed from food exporting to broad self-sufficiency, exporting mainly processed and value-added foods and importing an increasing number of basic commodities.

6 See also the 'European food declaration' by a coalition of nearly 200 NGOs, on 16 March 2010: 'The European Union must recognize and support the crucial role of sustainable farming in the food supply of the population. All people should have access to healthy, safe, and nutritious food. The ways in which we grow, distribute, prepare and eat food should celebrate Europe's cultural diversity, providing sustenance equitably and sustainably' (European Civil Society, 2010, www.europeanfooddeclaration.org, accessed 21 March 2011).

7 ARC is 'a common initiative of a European think tank working on agriculture policies called "Groupe de Bruges"'. ARC's views are supported by a wide range of European social movements and interest groups (see www.arc2020.eu, accessed 18 March 2011).

8 Some are fundamentally sceptical about the possibility that global food markets will contribute to sustainable food provision at all. For instance, Van der Ploeg (2010, p104) argues that 'the "world market" is an intrinsically unstable organizing principle. It constantly produces disequilibria, insecurity and turbulence (just as it is unable to create the required coordination between the production of biofuels and food).'

9 Since the 1980s, public investments in research and development for agriculture have substantially decreased, and technological innovations depend much more on privately funded research.

10 An example of the need for more knowledge-intensive research is the concentration on just a few crops and the neglect of 'orphan' crops. Orphan crops, such as millet and quinoa, have been ignored for a long time in research and development, but they could contribute substantially to increased agricultural production and by offering strengthened resilience.

11 In recent years, agriculture is increasingly being seen as part of a bio-based economy, where renewable inputs are produced for food, fuels and industry. See Langeveld et al (2010).

12 An example of such contestations is the use of GM crops. GM crops may have a role in the future of food, but ecologically integrated approaches – such as integrated pest management, minimum tillage, drip irrigation and integrated soil fertility management – often score higher in terms of resilience, as they put power in the hands of farmers rather than seed companies.

13 At present, there is a wider debate on steering technological development, and many thereby focus on how to increase knowledge and public control to prevent risks from happening. As Beck (1999) argues, however, it may be more advantageous to recognize that strict control of technology in global modernity is not possible because there are too many unknown and unknowable consequences of contemporary innovations that interact with humans and the environment. Our society is a global risk society, so when dealing with new technologies, not just experts' advice is needed, but broad participation and public debate as well, to balance the risks of social and environmental impacts (side effects).

## References

Andriesse, W., Giller, K., Jiggins, J., Löffler, H., Oosterveer, P. and Woodhill, J. (2007) *The Role of Agriculture in Achieving MDG1: A Review of the Leading Reports*, WUR, Wageningen

ARC (Agricultural and Rural Convention) (2010) *A Communication from Civil Society to the European Union Institutions on the Future Agricultural and Rural Policy*, ARC, Brussels

Beck, U. (1999) *World Risk Society*, Polity Press, Cambridge

Bieckmann, F. (2011) 'Reshuffling power', *The Broker*, vol 24, p2

Borlaug, N. (2000) 'Ending world hunger: The promise of biotechnology and the threat of antiscience zealotry', *Plant Physiology*, vol 124, no 2, pp487–490

*Bridges Weekly* (2010) 'European Union publishes long-awaited farm policy proposal', *Bridges Weekly Trade News Digest*, vol 14, no 41, 24 November, pp3–5, http://ictsd.org/downloads/bridgesweekly/bridgesweekly14-41.pdf (accessed 10 March 2011)

De Schutter, O. (2010) *Report Submitted by the Special Rapporteur on the Right to Food*, United Nations General Assembly, Geneva

EC (European Commission) (2010) *The Common Agricultural Policy after 2013: Public Debate*, European Commission, Brussels

European Civil Society (2010) *European Food Declaration*, European Civil Society, Brussels

Evans, A. (2009) *The Feeding of the Nine Billion: Global Food Security for the 21st Century*, The Royal Institute of International Affairs Chatham House, London

Foresight (2011) *The Future of Food and Farming: Challenges and Choices for Global Sustainability. Final Project Report*, The Government Office for Science, London

Freidberg, S. (2004) 'The ethical complex of corporate food power', *Environment and Planning A*, vol 22, pp513–531

Ghosh, J. (2010) 'The unnatural coupling: Food and global finance', *Journal of Agrarian Change*, vol 10, no 1, pp72–86

Hunt, D. and Lipton, M. (2011) *Green Revolutions for Sub-Saharan Africa?*, Chatham House, London

Lang, T. (2010) 'Crisis? What crisis? The normality of the current food crisis', *Journal of Agrarian Change*, vol 10, no 1, pp87–97

Lang, T., Barling, D. and Caraher, M. (2009) *Food Policy: Integrating Health, Environment and Society*, Oxford University Press, Oxford

Langeveld, H., Sanders, J. and Meeusen, M. (eds) (2010) *The Biobased Economy: Biofuels, Materials and Chemicals in the Post-oil Era*, Earthscan, London and Washington, DC

Matthews, A. (2010) *How Might the EU's Common Agricultural Policy Affect Trade and Development After 2013? An Analysis of the European Commission's November 2010 Communication*, ICTSD, Geneva

Ministère de l'Ecologie, de l'Énergie, du Développement Durable et de la Mer (2010) *Pour une politique agricole durable en 2013. Principes, architecture et éléments financiers*, Ministère de l'Ecologie, de l'Énergie, du Développement Durable et de la Mer, Paris

Murphy, S. (2006) *Concentrated Market Power and Agricultural Trade*, Heinrich Böll Foundation, MISEROR and the Wuppertal Institute for Climate, Environment and Energy, Berlin, Aachen and Wuppertal

Nelson, G., Rosegrant, M., Palazzo, A., Gray, I., Ingersoll, C., Robertson, R., Tokgoz, S., Zhu, T., Sulser, T., Ringler, C., Msangi, S. and You, L. (2010) *Food Security, Farming, and Climate Change to 2050: Scenarios, Results, Policy Options*, IFPRI, Washington, DC

*New York Times* (2009) 'Experts worry as population and hunger grow', *New York Times*, 22 October

Timmer, P. (2008) 'Food policy in the era of supermarkets: What's different?', in E. McCullough, P. Pingali and K. Stamoulis (eds) *The Transformation of Agri-Food Systems: Globalization, Supply Chains and Smallholder Farmers*, Earthscan, London and Sterling, pp67–86

Van der Ploeg, J. (2010) 'The food crisis, industrialized farming and the imperial regime', *Journal of Agrarian Change*, vol 10, no 1, pp98–106

Vellema, S. and Danse, M. (2007) *Innovation and Development: Institutional Perspectives on Technological Change in Agri-food Chains*, DLO, Wageningen

# Index

Page numbers in *italic* refer to boxes, tables and figures